The End of the Empty Organism

The End of the Empty Organism

NEUROBIOLOGY AND THE SCIENCES OF HUMAN ACTION

Elliott White

HUMAN EVOLUTION, BEHAVIOR, AND INTELLIGENCE
Seymour Itzkoff, *Series Editor*

PRAEGER

Westport, Connecticut
London

Library of Congress Cataloging-in-Publication Data

White, Elliott.
   The end of the empty organism : neurobiology and the sciences of
human action / Elliott White.
      p.   cm. — (Human evolution, behavior, and intelligence, ISSN 1063–2158)
   Includes bibliographical references and index.
   ISBN 0–275–94368–2 (alk. paper)
   1. Neurobiology—Social aspects.   2. Neurobiology—Philosophy.
I. Title.   II. Title: Neurobiology and the sciences of human action.
III. Series.
   [DNLM: 1. Behavior—physiology.   2. Neurobiology.
3. Neuropsychology.   WL 103 W583e]
QP356.W44   1992
612.8′01—dc20
DNLM/DLC
for Library of Congress   92–3571

British Library Cataloguing in Publication Data is available.

Library of Congress Catalog Card Number: 92–3571
ISBN: 0–275–94368–2
ISSN: 1063–2158

First published in 1992

Praeger Publishers, 88 Post Road West, Westport, CT 06881
An imprint of Greenwood Publishing Group, Inc.

Printed in the United States of America

∞™

The paper used in this book complies with the
Permanent Paper Standard issued by the National
Information Standards Organization (Z39.48–1984).

10 9 8 7 6 5 4 3 2 1

To the memory of my mother and father

New concepts of mind and brain have helped to bring a
profound change of paradigm. . . . Based on the shift to a
new causal interpretation of conscious experience, it has
overthrown long-accepted ideas of twentieth-century
scientific materialism. The repercussions of the new view
of consciousness pervade the entire structure of the
scientific approach to human nature.

—R. W. Sperry (1983)

Contents

Figures

Preface

I have written this work as a political scientist who became interested in the ramifications of the life sciences in general and of neurobiology in particular for my area of study; but as I believe that the work has a broader applicability to the social sciences more generally, I have not hesitated to suggest its larger implications.

In this endeavor, I have benefited from my colleagues in political science who also have been interested in the life sciences and their relevance, including Joe Losco, Roger Masters, Glen Schubert, John Wahlke, and the late Tom Weigele.

I thank the editor in chief of Paideia Publishers who referred me to the editor of this series on human evolution, Seymour W. Itzkoff, whose prompt and supportive reading of this manuscript has in turn facilitated its publication. I am greatly indebted to Gloria Basmajian who, as in the past, has somehow succeeded in deciphering my handwriting and transferring it by means of the word processor into presentable typed copy.

And through it all, I have been sustained by my wife Arleen.

1

The Empty Organism

... the ideal would be ultimately to get away from conscious dimensions to physical dimensions, to the happy monism of the scientific heaven. ... we are not yet ready to give up the conscious dimensions. We need them now, but I think we are already seeing how it can come about that we shall eventually be able to do without them.

—Edwin Boring (1933)

A science of human life that ignores the brain is akin to a study of the solar system that leaves out the sun. This statement assumes a heliocentric solar system and autonomous human organisms. For just as the sun is relegated to a peripheral position in a Ptolemaic or geocentric system, so the human brain lacks critical importance from a mechanistic perspective whereby man's thought becomes largely the product of external or internal forces, whether of environmental, genetic, or physicochemical origin. Thus human thinking and consciousness are, from this latter view, at most epiphenomenal, with no causative force of their own. It follows that an "empty organism" approach not only ensues; it also becomes wholly justified. For why bother to consider the biological individual if his thought and action are all determined by more critical factors that lie elsewhere? In this chapter, accordingly, I wish to examine several reductionist models of the human brain and their differing implications for social science. In these models consciousness becomes reduced to environmental, genetic, or physicochemical factors.

AN ENVIRONMENTALIST MODEL

For a behaviorist psychologist, environmental factors determine human thought and action. As a result, the central nervous system can largely be ignored in the study of human life.

The fact that the neurosciences were so clearly in their infancy during the development of behaviorism meant that they provided little in the way of a solid foundation on which a psychology might build. "As a matter of fact we do not yet know enough about the functioning of brain and spinal cord," John B. Watson (1924, p. 43) was to note, "to draw diagrams about their functions." Skinner (1979, p. 68) recalled that early in his career (in 1928) he faulted the neurophysiology of C. S. Sherrington, the outstanding English neuroscientist of his time, on the grounds that Sherrington's description of the properties of the synapse, the point of contact between nerve cells, was not empirically based: "I argued that he had never seen a synapse in action and the properties could be defined operationally by referring to behavior and environment without mentioning the nervous system." Skinner (1979, p. 68; emphasis in the original) went on to conclude that he "rejected Sherrington's physiology" because he "wanted a science of *behavior*." Here Skinner remained consistent with Watson's first commandment of behaviorism (1924, p. 6): "Let us limit ourselves to things that can be observed, and formulate laws concerning only those things."

Watson's injunction in effect guided Skinner throughout his long career. For in one of his later works, Skinner (1976, pp. 234, 240) conceded that while "the parts of the nervous system spoken of by early physiologists were, as we have seen, largely a matter of inference, the classical example being the synapse of Sherrington's *The Integrative Action of the Nervous System*," now "inference has yielded to direct observation as instruments and methods have been improved, and with great gains for physiology." Nonetheless, Skinner clearly did not believe that neurophysiology had become sufficiently advanced to serve as a direct basis for psychological study. He therefore raised the question: "Will a model of the nervous system not serve until physiology is more advanced?" And his answer was no—because the " 'thought processes' [that] are real enough at the level of behavior [are] merely questionable metaphors then moved inside" (cf. Skinner 1983, pp. 279, 367).

The behaviorist insistence on the observability of his subject matter limits that subject matter basically to two factors: the environment and the behavior of the organism. The behaviorist proceeds to assume that the former (the environment) determines the latter (the behavior). Such an approach hardly emphasizes an independent role for the central nervous system. As Watson (1919, p. 20) writes, "the physiologist qua physiologist knows nothing of the total situations in the daily life of an individual that shapes his action and conduct." Skinner (1971, p. 180) nonetheless maintained that in shifting control "to the observable environment we do not leave an empty organism": "A great deal goes on inside the skin, and physiology will eventually tell more about it. It will explain why behavior is indeed related to the antecedent events of which it can be shown to be a function." In other words, both the functioning of the brain together with the behavior of the organism become, from this viewpoint, the product of external environmental contingencies. Subjective states of mind may exist, moreover, but they are at best epiphenomenal in character. They have no independent causal status (see Koestler 1967, ch. 1).

If Skinner should be proven right, social science becomes an extension of behaviorist psychology; for then both an individual's brain and his social behavior will be explicated in environmentalist terms. Methodologically, the focus of research becomes the external behavior of the individual. As Skinner (1976, pp. 265–66) observed, "behaviorism could perhaps be called reductionistic when it discusses the social sciences. It reduces social processes to the behavior of individuals."

Insofar as the social sciences refer to themselves as behavioral sciences and stress the influence of environmental factors, they share the basic outlook of the behaviorist. But clearly they did not go far enough to satisfy Skinner (1971, p. 22): "One difficulty is that almost all of what is called behavioral science continues to trace behavior to states of mind, feelings, traits of characters, human nature and so on."

But before adopting Skinner's behaviorism, social scientists might note that Skinner simply assumed that neurobiology would come to validate his basic approach. He assumed what must remain, as Chomsky (1971) points out, a critically open question. In the absence of definitive evidence, however, the social scientist might nonetheless ask what current neurobiological theory has to say concerning the question: Is the functioning of the central nervous system in fact determined by environmental forces?

Now if the writings of leading neuroscientists were to be taken as holy writ, one could cite scripture at some length regarding the invalidity of an environmentalist model of the brain. As the late E. H. Lenneberg (1973, p. 128) stated: "A nervous system is not like a trumpet into which the environment can blow and produce a tune. Brains are not passive conveyers of information. . . . Psychologically important stimuli trigger and shape behavior, but the stimuli are not the architect of the principle by which behavior operates; nor is behavior a transform of the brain's input." Similarly, Granit (1977, p. 67) asserts that "the central nervous system by no means behaves as a passive receiver of input. . . . It selects its information actively by processing it in the periphery, at the subcortical level, and within the cortex, where the ultimate selection takes place with the aid of consciousness." Therefore, as McGuiness and Pribram (1980, p. 101) point out, the interaction between organisms and their environment is "not one-sided," for organisms are "spontaneously active."

On "the evidence available today," Sperry could write by the mid–1960s (1965, p. 91), "we should renounce, along with other aspects of the behaviorist, materialist approach . . . the old Pavlovian-Watsonian conditioned-reflex theory of the psyche, with its radical environmentalism that used to tell us that literally 99 percent of human nature and mind is a product of experience and training." Responding directly to the behaviorism of Skinner, Eccles (1973, p. 89) concluded that "the behavioristic dogma of Skinner is that investigations on the pattern of stimulus-response-reinforcement will eventually lead to a complete explanation not only of animal but also of human behavior, and to its complete control by operant conditioning, as witness his . . . *Beyond Freedom and Dignity*. I reject this philosophy because it resolves the brain-mind problem by ignoring both brain and mind; the former is safely and inviolably enclosed in the black box, and the latter is as ineffective as a fantasy."

Klivington (1989, p. 176) summarizes our current understanding of the brain as follows:

We have seen that experience-dependent modifications of connectivity depend to a critical extent also on internally generated gating signals. Thus, large neuronal arrays participate in the decision whether a particular activation pattern should lead to long-lasting modifications of circuitry. Experience-dependent self-organization, therefore, has little in common with passive instruction of a *tabula rasa*. Rather, the developing brain appears as a highly active and primarily self-contained system that, at birth, already possesses substantial knowledge about the structure of the world to which it is going to adapt itself.

Indeed, Gallistel (1989, pp. 184–89), in contrasting a passive "Lockian" with an active "Leibnizian" view of the brain, argues that in the animal world generally, the active view is the better fit. His review of the experimental data supports "the view that natural selection has shaped brains in such a way that there exists the kind of brain-world parallelism of formal structure that Leibniz seems to have envisaged." An environmental reductionism may attempt to build a hard systematic science; but it may produce an invalid one.

A GENETIC MODEL

The emergence of the biobehavioral sciences would, one might presume, suggest an integral role for the biological organism. Yet such has thus far not been the case. As Corning (1983) observes, no biological approach to the study of human life is in fact at present biological in the sense of directly treating the human organism.

Sociobiology is not exempt from this critical observation. Thus Schubert (1981b, p. 193) refers to sociobiological, or as he calls him, "Bionic man," as "equally devoid of flesh (and bones too, for that matter!)." Let us examine sociobiological theory more closely, with special emphasis on the formulations of Edward O. Wilson, to discern the role that it gives to the biological organism.

Sociobiology traces its intellectual ancestry to Charles Darwin, but its name, its systematic exposition, and its wider and controversial recognition awaited Wilson's publication in 1975 of the monumental synthesis by that name. How much the term, if not the approach, is uniquely Wilson's can be seen by the fact that in the following year Richard Dawkins's *The Selfish Gene*, covering some of the same ground, mentions the term *sociobiology* and, for that matter, E. O. Wilson only when referring to the title of Wilson's volume—and that by way of a critique.

The most important concept that sociobiology takes from Darwin is that of natural selection: the notion that the environment favors the survival and successful reproduction of some organisms as against others—with the presumption that the individual differences that are involved have an important genetic component. The characteristics that may be selected for and that may have a genetic basis include the behavior of the organism; thus Darwin, in explicitly recognizing this possibility, became in effect the father of ethology, the study of animal behavior within the context of natural selection and the evolutionary process.

The recognition that the gene was the particulate unit of heredity was Mendel's and not Darwin's and did not become incorporated into Darwinian evolution until the advent of the present century. This marriage of Mendel and Darwin, the modern synthetic theory of evolution, made possible as well the later union of ethology with the field of population genetics, that is, the study of the statistical frequencies over time of the genes in animal populations and the change of those frequencies over time, the latter development encompassing the evolutionary process.

The immediate cause of the union of ethology and population genetics—out of which came sociobiology—was the publication in 1962 of Wynne-Edwards's *Animal Dispersal in Relation to Social Behavior*. Wynne-Edwards argued that individual animals reduced their own birthrates for the good of the group as a whole. Thus Wynne-Edwards argued for group selection as the mechanism for evolutionary change. As Dawkins (1976, p. 8) characterizes it, "a group, such as a species or a population within a species, whose individual members are prepared to sacrifice themselves for the welfare of the group, may be less likely to go extinct than a rival group whose individual members place their own selfish interests first. Therefore, the world becomes populated mainly by groups consisting of self-sacrificing individuals."

The most obvious problem for group selection theory was in showing how organisms that restricted their reproduction would be selected for against organisms that did not. Hence the question that, according to Wilson (1975, p. 3), constitutes "the central theoretical problem of socio-biology: how can altruism, which by definition reduces personal fitness, possibly evolve by natural selection?"

Almost in the very nature of things, would not the more selfish organism also more successfully reproduce? If so, would not therefore individual selection, not group selection, constitute the basic mechanism of the ev-olutionary process? And yet even opponents of group selection theory had to admit that some behavior appears self-sacrificing or altruistic, most clearly in the case of the mammalian mother and her concern for her offspring but also recognizably in such cases as warning calls by one animal to others in a group. How does one explain such apparent examples of altruistic behavior while at the same time rejecting group selection?

There is one possible response, which indeed, if valid, constitutes the distinctive contribution of the new sociobiological literature to contem-porary biology. Those rejecting the theory of group selection substituted the gene as against either the group or the individual as the unit of natural selection. The evolutionary process thus comes to be viewed from the perspective of the gene, and all life comes to be understood as the result of the apparent effort to survive of the genetic units that are contained on the long nucleotide chain of the DNA molecule.

The "selfish" gene makes possible, however, the altruistic individual. To understand this possibility, we should understand first what Barash (1977, pp. 63, 81) terms "the Central Theorem of Sociobiology": "When any behavior under study reflects some component of genotype [i.e., is genetically influenced], animals should behave so as to maximize their inclusive fitness," inclusive fitness being defined as "their genetic repre-sentation in succeeding generations, including other relatives in addition to offspring."

To understand "the central theorem," one must keep in mind that from the perspective of the gene, the survival of the gene—not the individual and not the group—is paramount. If, however, a number of individuals are closely related, they necessarily share a number of genes in common. A child, for example, receives half of his genes from each of his parents and will, on the average, share half of his genes with each of his siblings. Thus if a parent or even one of the children in a four-member family sacrificed his life to save the remaining three members (say, from a burning house), the sacrifice, from the perspective of the gene, would be more than justified: the cost-benefit ratio is a net plus since more of the same genes survive than would be the case if no self-sacrificing act on the individual level were attempted. The individual is altruistic, the genes remain selfish. (Presumably, therefore, on a strict cost-benefit analysis, it would not "pay" for a mother to sacrifice herself for "only" one of her children.)

As we are constantly reminded in the sociobiological literature, when one says that the genes attempt to survive or that organisms are altruistic, no conscious acts are to be inferred: rather genes and organisms act as if they might be attempting to survive (or to sacrifice themselves in order that others may survive), natural selection being the mechanism that gives the appearance of deliberate choice.

Individual altruistic behavior can be explained, then, by "kin selection." This concept, which originated in the work of the British population geneticist W. D. Hamilton in the mid–1960s, refers to the selection of genes because of their effect in favoring the reproductive success of relatives, and according to Barash (1977, pp. 84–85) "is of enormous explanatory power and is basic to the intellectual momentum of sociobiology."

In its most mechanistic and reductionistic formulation, sociobiology allows no independent role to the organism. Dawkins (1976, p. 157) puts it starkly: "A body is really a machine blindly programmed by its selfish genes." Wilson appears to be in substantial agreement. Where Dawkins (1976, p. 25) succinctly says, "A body is the gene's way of preserving the genes unaltered," Wilson (1975, p. 3) asserts, paraphrasing Samuel Butler, that "the organism is only DNA's way of making more DNA." Therefore, as Dawkins (1976, p. ix) asserts, we humans "are survival machines—robot vehicles blindly programmed to preserve the selfish molecule known as genes."

Yet Wilson is also explicitly mindful of the potential role that the brain sciences will come to play in the study of man. Indeed, in the very first chapter of his *Sociobiology* (1975, p. 6), he reserves a more or less coequal role to "neurophysiology" in a larger emerging "behavioral biology," that is, the "scientific study of all aspects of behavior." In other words, it is not sociobiology alone, but rather its union with neurophysiology that will result in a more comprehensive behavioral biology. Wilson proceeds to ignore this formulation throughout the remainder of *Sociobiology* and throughout the whole of his sequel *On Human Nature*, failing significantly even to refer explicitly to behavioral biology.

An emerging behavioral biology will, in any case, according to Wilson, come to "cannibalize" the existing social sciences. If indeed the brain is blindly programmed by the genes, whose "interests" it serves exclusively, then both social science and neurobiological theory become isomorphic with sociobiology.

Nonetheless, on the very last page of his *Sociobiology* (1975, p. 557) Wilson concludes that the development of "fundamental theory in sociology must await a full, neuronal explanation of the human brain. . . . Having cannibalized psychology, the new neurobiology will yield an enduring set of first principles for sociobiology." Having reserved such a critical future role for an emerging neurobiology, Wilson would appear to have rendered his own sociobiological approach at the least premature.

What Wilson does is to anticipate the future source of neurobiology by assuming it to be fully compatible with current sociobiological theory. Wilson begins *Sociobiology* (1975, p. 3), for example, with the assertion that "self-knowledge is constrained and shaped by the emotional control centers in the hypothalamus and limbic system of the brain," which "flood our consciousness with all the emotions . . . that are consulted by ethical philosophers who wish to intuit the standards of good and evil"; and these "control centers" having evolved through natural selection, must therefore, in their functioning, manifest the operation of sociobiological principles. As Wilson directly puts it in *On Human Nature* (1978, p. 2), "the brain exists because it promotes the survival and multiplication of the genes that direct its assembly. The human mind is a device for survival and reproduction, and reason is just one of its various techniques." The lower centers of the brain thus control the higher, or as Wilson (1978, p. 57) approvingly quotes David Hume, "reason is the slave of the passions."

Wilson (1978, p. 75) not only anticipates future neurobiological findings; he also evaluates present models of the human brain by how closely they cohere to his sociobiological principles. Thus if too simple a neurological model is to be avoided, so also "too complex a neurological model can lead back to a vitalistic metaphysics." Since the latter model can lead back to conclusions theoretically unacceptable to Wilson, "a compromise solution" is advocated.

Wilson, then, in anticipating future neurobiological research to be fully compatible with his theoretical approach, is in this respect similar to Skinner (Fuller 1978; and Frankel 1979, pp. 43–45). Wilson and Skinner clearly have differing expectations of the findings of future neurophysiological research and both cannot be right—although both could be wrong (see Searle 1978; Grene 1978).

Both Wilson and Skinner are in fundamental agreement, however, in rejecting mentalist models of the brain. Wilson (in Gregory et al. 1978, p. 9) views "mentalism" to be "the strongest redoubt of counterbiology." While it may be more accurate to say that mentalism is "the strongest redoubt" of a "countersociobiology," Wilson (1978, p. 195) insists that "the mind will be more precisely explained as an epiphenomenon of the neuronal machinery of the brain." As the late C. H. Waddington (1978, p. 257) critically remarked in his review of *Sociobiology*, "Is it not surprising that in a book of 700 large pages about social behavior there is no explicit mention whatever of mentality? In the index, covering more than thirty pages of three columns each, there is no mention of mind, mentality, purpose, goal, aim, or any world of similar connotations."

Wilson's scientific materialism, which treats the mind as epiphenomenon, is therefore on this basic point similar to behaviorism, which Wilson (1978, p. 65) explicitly acknowledges: "The central idea of the philosophy of behaviorism, that behavior and the mind have an entirely materialist basis subject to experimental analysis, is fundamentally sound." Unlike behaviorism, sociobiology stresses the biological basis of all behavior; but like behaviorism, the most abstract and sophisticated modes of thought and action are understood by reference to factors of a lower, presumably more basic nature.

Searle (1978, p. 168) accordingly denies, on these grounds, that "there is a great difference between sociobiology and old-fashioned behaviorism": "Both are deterministic about all human behavior in the strict sense that they see each action as caused by antecedently sufficient conditions, and they both reject any mentalism in the explanation of behavior. The only inference is in the relative weight that they attach to genetic constraints—the behaviorist tends to emphasize the causal role of operant conditioning in the determination of behavior and the sociobiologist tends to emphasize the causal role of the genotype" (cf. Skinner 1983, p. 384). Both behavioral and biobehavioral science alike then operate upon the basis of an "empty organism" in which the functioning of the brain is assumed to reflect the same extracerebral influences, whether environmental or biological, that the approach in question sees as determining behavior.

The very real possibility exists, however, that such efforts to bind future brain research to present sociobiological (or behaviorist) theory is premature and unrealistic. Even sociobiological theory may not be fixed at this point. Wilson himself (in Davis and Flaherty 1976, p. 11) has raised the possibility that "a wholly new picture of organic evolution may be about to emerge, and that this may be one of the turning points in the history of science." If so, Wilson's present speculations on the science of the brain would become quickly dated. But his speculations may well become dated in any case—for the neurosciences will no doubt develop in their own way and, to paraphrase Wilson, "a wholly new picture" of the human brain "may be about to emerge," and "this may be one of the turning points in the history of science."

More recently Wilson, in conjunction with Lumsden (Lumsden and Wilson 1981), has attempted an explicit synthesis of "genes, mind and culture." But in its concern to build a quantitative, systematic, and predictive science, unique individuals become overshadowed. The "mind" in the title of the work is found only in populations as a statistical entity; it has no personal embodiment. As Corning (1983, p. 123) so graphically observes of their work: "The mind cannot be likened to a random assortment of decision rules, like so many pickup sticks. Despite their claims to the contrary, Lumsden and Wilson have no theory of the mind qua system. As Gertrude Stein said of Oakland, there is no there there."[1]

A similar point is made by Kitcher (1985) in his massive critique of sociobiology in general and Wilson's work in particular. Although perhaps too eager to discredit the operation of genetic influences, especially in human affairs, Kitcher (1985, pp. 417–34) legitimately questions what he terms "the Hypothalamic Imperative," the "most ambitious" of "socio-biological adventures in philosophy," that is the reduction of human ethics and self-knowledge to activity of "the emotional control centers in the hypothalamus and limbic system of the brain." Kitcher (1985, p. 431) in-dicates how disembodied Wilson's analysis is in concrete biological terms when he discusses Wilson's advocacy of "universal human rights" on the grounds that our society was built "on the mammalian plan." "*What* mam-malian plan?" is Kitcher's response (1985, p. 431; emphasis in the original). Mammalian societies, Kitcher proceeds to observe, "are highly diverse." Thus *pace* Wilson, no clear "mammalian imperative" emerges.

In any case, while awaiting any further departures in sociobiological or neurobiological theory, how is one to evaluate the neurobiological as-sumptions made within current sociobiological theory? An observer, nec-essarily relying on the writing of leading neuroscientists, must first of all report that a genetic reductionism appears no more valid than an environ-mentalist one. The brain, writes Eccles (1978, p. 175), "is not a fixed action structure. At the higher levels of the brain, modifiability is the essence of its performance, as is evidenced by learning and memory": Our "personal uniqueness and all aspects of its associated experiences are dependent upon the brain; yet it is not entirely dependent on the genetic instructions that built the brain in the manner that is still only vaguely understood."

An emphasis on the role of the genes tends to downplay the importance of the developmental process, including the role played by environmental factors. As Stent (1976, p. 190) writes, "The deep biological problem in want of a solution is not to find out how genes determine behavior but to work out the general rules under which the nervous system is produced. The horizon of the problem lies far beyond the genes and encompasses mainly the context under which the molecular meaning of genetic infor-mation gives rise to its epigenetic meaning."

A phylogenetic reductionism is also questionable. MacLean (1973, p. 58), in his "triune concept of the brain and behavior," does not like Wilson simply reduce the higher and later levels of the brain to the earlier and more primitive. The "neomammalian" brain does not merely serve the "reptilian," even if the first layer may still influence the latest. Citing an earlier work of his, MacLean writes that "in the complex organization of the phylogenetically old and new structures ' . . . we presumably have a neural ladder, a visionary ladder, for ascending from the most primitive sexual feeling to the highest level of altruistic sentiments.' . . . it is possible that these large evolving territories of the brain are incapable of being brought into full operation until the hormonal changes of adolescence occur. If so, it would weight heavily against the claims of those who contend that the personality is fully developed and rigid by adolescence, if indeed not by the age of five and six." And certainly not, therefore, by birth or conception. Schmitt (Schmitt et al. 1982, p. 2), moreover, points out that the essence of a theoretical statement by Vernon Mountcastle (1978) is that "although phylogenetically older parts of the brain may play a significant role, the key to any fruitful theories of higher brain function must be the unique structure and properties of the cerebral cortex."

That much can transpire between the initial genetic program and the resulting individual phenotype is clearly shown by Gazzaniga (1988, pp. 234–37). He reports a case of identical twins raised apart, both by stepparents. One set of parents were both professors, with a "rational approach to the everyday," quite exacting and achievement oriented. The other set were less well educated, more casual and less demanding. It turned out that both twins had Huntington's disease, each having inherited "the same exact gene" for it. In this disorder, however, the symptoms can be diametrically opposed: some victims "experience a slowness of movement" resembling a symptom of Parkinson's disease; while for others "it is just the opposite; their movement became rapid." As it developed, one twin "became hyperkinetic and the other was Parkinsonian." Gazzaniga (1988, pp. 235–36; cf. Plomin 1990, pp. 186–87) argues as follows:

Our idealized example of these two women points out the very real possibility that mind influences can well affect the genetic imperative. While the straight genetic view would tend to explain the variation in symptoms by citing a difference in the penetrance of the genes or the probability that several other genes had interacted in certain ways, it is just as likely that the mind states of the women were preparing their bodies to accept two different kinds of genetic information. In other genetically driven diseases, the concordance rate is nowhere near as high, thereby leaving even greater room for mind influences as well as other epigenetic influences. The suggestion here is that studies of twins reveal that the mind may be playing a major role in the response rates for disease.

The view of human self-awareness as epiphenomenal is, finally, a curious one for a self-proclaimed evolutionary theory to take. For if self-awareness has indeed evolved and yet serves no independently important function, why has it evolved in the first place (cf. Granit 1977; Searle 1978)? Since sociobiologists stress the adaptive function of physical and psychological traits, why is the most unique of human characteristics relegated to epiphenomenal status? Natural selection implies, Symons (1979, p. 89) states, that "were mind functionless it would long ago have withered away, not become more elaborate, as it obviously has."

The validity of key sociobiological premises thus remains still very much in doubt. The social scientist can hardly ignore an emerging discipline that would render his own extinct; but he may nonetheless be justified in exercising caution before leaping uncritically into its clutches.

A PHYSICOCHEMICAL MODEL

Until recently at least, it had been the expectation of many biologists to explain all of life solely by reference to the atoms and molecules that constitute it. Thus following his codiscovery of the molecular structure of the gene, Crick (1966, p. 10, emphasis in the original) wrote that "the ultimate aim of the modern movement in biology is in fact to explain *all* biology in terms of physics and chemistry." Crick (1966, p. 13) acknowledges that "a biological system can be regarded as a hierarchy of levels of organization, the 'whole' of one level being the parts of the next." He nonetheless contends (1966, p. 14) that "eventually one may hope to have the whole of biology 'explained' in terms of the level below it, and so on right down to the atomic level."

Presumably, therefore, the human brain itself would not be exempt from this atomic reductionism—although Crick (1966, p. 74) concedes that in understanding ourselves, "how our brains work and why we are conscious," we "need to grasp the behavior of the organism as a whole, and all the complex interactions that go into making that behavior." To the extent, however, that the focus of a physicochemical model remains atomistic and reductionist, to that extent the organism becomes mechanized; it may not be "empty," inasmuch as it is comprised of its critically important component parts; but it becomes more a machine than an autonomous organism. Crick (1966, pp. 44–84), for example, while clearly aware of the critical differences between brains and computers is also clearly enamored with the possibility of an artificial intelligence. (It is far from incidental that the image of the individual as a machine is explicit in the reductionist approaches already discussed. Thus Skinner [1953, p. 45] subtitled one of his chapter sections "Man a Machine" and Wilson [1978, pp. 55, 56] refers to the newborn infant as a "marvelous robot" that is "wired with awesome precision.")

Crick (1966, p. 74) acknowledges that "the study of the nervous system is going to have to continue for a considerable period before it can answer those questions about ourselves which most interest us." If the time arrives, however, when the brain can be explained exclusively by reference to its parts, the larger implications for social science will be impossible to ignore. According to Olby (1970, p. 970), a molecular biologist "would not claim that a piece of Shakespeare's writing is reducible to physics and chemistry though he would claim that the thought processes which generate these will be explicable in terms of those sciences." The latter claim is nonetheless a striking one; and it is not clear how molecular biology, on this view, stops short of accounting for the products of Shakespeare's (or anyone's) thought, if it indeed accounts for the thought itself. In any case, psychology and social science would virtually become an extension of an all-embracing molecular biology. Schubert's advocacy of a curriculum in which future social scientists become well grounded in the "hard" physical and biological sciences (1973) may not be without considerable merit whatever model of the brain becomes accepted; but with this model such a curriculum becomes vindicated perhaps beyond Schubert's most fervent wishes.

Gardner (1985), writing as cognitive scientist, strongly defends its autonomy as against a neuroscientific reduction. Raising the question "will neuroscience devour cognitive science?" Gardner (1985, pp. 286–87) replies that although perhaps a majority of neuroscientists might respond in the affirmative, he would paraphrase Wittgenstein—"one can know every brain connection involved in a concept formation, but that won't help one bit in understanding what a concept is." In a discussion of language that is relevant to Olby's reference to Shakespeare's writing, Gardner contends that only a linguistically trained observer can perceive subtle but critical distinctions in diagnosing an aphasic condition; and he concludes that as a result, "it is now recognized that an aphasia research team should include psychologists and linguists as well as neuroscientists."

The jury, of course, is still out on whether, as Churchland (1988, p. 273) puts it, "mental phenomena might be reducible to neurobiological phenomenon." Writing as a "neurophilosopher," she appears to believe such a reduction to be possible, even probable, but perhaps not necessarily "in the classical sense," that is to say, as reduction has been conventionally understood to subsume the higher by the lower. Rather Churchland (1988, p. 301) suggests that we may "have to re-think our pre-scientific intuitions concerning what it is to be conscious." Even if Churchland is right, this suggestion only begs the question—or perhaps more accurately, indicates that the question is still premature.

Indeed, one possibility remains that the question in some sense will always be premature. The Italian neurologist Bisiach (1988, p. 117) observes that questions that relate to consciousness, insofar as they refer narrowly to inner subjective phenomenal mental states may "remain forever unanswerable." The study of consciousness, however defined, is further complicated, Bisiach continues, by the fact that it is "basically multidisciplinary and the involved parties differ considerably in background area of expertise, and personal as well as group bias."

If in theory a wholly reductionistic understanding of the brain and of our behavior were possible, however, in practice it might not be. P. W. Bridgman, the Nobel Prize–winning physicist whose operational approach has been so influential in psychology and the behavioral sciences, as well as in the natural sciences, argued (1959, ch. 6) that operational understanding of consciousness in atomistic terms is technically impossible. Bridgman (1959, p. 201) examined what he terms "the thesis of sufficiency of atomic analysis," which he formulated as follows: "Given a complete description in physical terms of any organism, then there is nothing more to give, in the sense that all the present behavior of the organism and its future behaviors in a completely specified environment are fixed." At the time of this writing—the late 1950s—Bridgman did not doubt that most biologists and psychologists would accept this thesis, only adding that most of the latter would state that "the most important part of the atomic specification is concerned with the atoms of the nervous system and in particular with the brain." Bridgman notes the wide scientific appeal of a method that permits one to treat in objective and operational terms areas of human life that are otherwise quite subjective and unaccessible—our dreams, our subconsciousness, and our consciousness.

As desirable as such an operational approach might be, Bridgman (1959, pp. 211–12; cf. Harth 1982, p. 103) argues that its implementation is technically impossible. First, "there is not room enough in space for all the instruments that would be required to measure the position of all the atoms." Second, any attempt to reduce the number of atoms to be measured by some sampling technique will likely fail because "organic systems are not homogeneous over as wide ranges as inorganic systems." Third, the translation of the atoms and their positions into mental states would be imposing, to say the least, when one considers "that it would take something of the order of 10^{14} lifetimes to even mention all the atoms, one by one, to say nothing of mentioning all their combinations, which according to supposition correspond to my different mental states." (It is of interest that Bridgman [1959, p. 216] reports having had long discussions with Skinner on the subject; and since Skinner [1979, pp. 67, 94] mentions his exposure to Bridgman's thought beginning in 1928, it is possible that it might have influenced his adverse reaction to his colleague E. O. Boring's physicalist approach to consciousness, which emerged in 1932; both Skinner [1976, p. 235] and Bridgman [1959, p. 211] also refer in effect to the application of the Heisenberg principle to the measurement of the central nervous system: the attempt to measure it will also disturb it; and, as Bridgman further notes, following Bohr, such an attempt may even kill the organism.)

Unlike Skinner (1979, p. 94) who was infuriated with Boring's physicalist approach for its return to physiology and for its "refusal to recognize the possibility of a science of behavior," Wilson (or other sociobiological theorists) have no necessary quarrel with a physicochemical focus on the brain. (Skinner, of course, does not deny the validity of such a focus; it is the focus itself—the emphasis away from external behavior—that constitutes the problem). As Rosenberg (1980, pp. 198–99) puts it succinctly: "the sociobiologist is a mechanist." Rosenberg sees both "the physicalist and the sociobiologist" in sympathy with "the mechanist's hoped-for neurophysiological theory," which "will cite the neurophysiological conditions which are causally sufficient for each and every internal movement of a human being, presumably by appeal to laws that are not restricted in their applicability to *Homo Sapiens*."

The critical question, however, remains: Is such a mechanistic, materialistic model of the brain in fact valid? As a working neuroscientist, Eccles (1974, p. 104) acknowledges that "even at the level of the cerebral cortex of animals and man it is essential to investigate the responses of nerve cells in all their patterned complexity and in all their learned responses on the basic postulate that it is all explicable in terms of biophysics, biochemistry, neurochemistry, molecular neurobiology, etc." Eccles goes on to state, however, "Yet at the same time it must be recognized that reductionism fails when confronted by the brain-mind problem."

Inasmuch as the following chapter will elaborate Eccles's contention, I do not pursue it here. Rather, I wish to call attention to Crick's more recent experience as a self-acknowledged "newcomer to neurobiology" and his subsequent realization (1979, p. 222) that "the brain is clearly so complex that the chances of being able to predict its behavior solely from a study of its parts is too remote to consider." Crick appears, in other words, to back away from his earlier more thoroughgoing reductionism.

Paradoxically, the complexity of the brain that Crick now so clearly recognizes appears at once to invite, even demand, the attention of psychologists and social scientists—and to repel it. On the one hand, Crick (1979, p. 222) points out that this "complexity also warns us that the black-box approach of pure psychology will have to be lucky if it is not to bog down." Overly simplistic assumptions concerning the functioning of the brain are likely to be shown invalid, and therefore psychology and social science "must combine the study of behavior with parallel studies of the inside of the brain."

On the other hand, the complexity of the brain makes premature any synthesis of neuroscience and social science. Crick (1979, p. 232), while taking note of the "interesting and exciting new work" that has been done on the brain, nonetheless concludes: "It is only when we reflect on how intricate the entire system is and how complicated the many different operations it has to perform are ... that we realize that we have a long way to go." Kurczman and Eccles (1972a) on the first page of their introduction to their volume on "brain and human behavior" were stuck with the "disconcerting idea" that the "volume may be premature by about 200 years," for "we are still at the primitive stage in the development of our knowledge of 'brain and behavior.' " If the future course of the neurosciences will eventually relate in some comprehensive way to human thought and action, that day, alas, may take centuries to arrive. Yet Crick's observation (1979, p. 32) appears irrefutable: "There is no scientific study more vital to man than the study of his own brain."

What, then, is the social scientist to do? Four choices are, I believe, open to him. First, he can simply ignore (or try to ignore) the problem. Second, he can assume future findings in neurobiology will fit his preexisting theoretical framework, just as Wilson and Skinner have done. Third, he can attempt to incorporate the technical knowledge now emerging from the neurosciences into current social science theory, perhaps qualifying the latter as it proves necessary. Finally, he can try to ascertain the more basic views of the brain that are emerging based upon the work of the leading authorities within the field. Let us briefly consider each of these alternatives in turn.

The social scientist has already shown himself to be adept in adopting the first solution: that of ignoring a field that appears both remote and complex. Only within the past decade or so have the emerging fields of ethology and behavior genetics been related to the social sciences (for political science, as an example, consult the comprehensive bibliography compiled by Somit, Peterson, Richardson, and Goldfischer [1980]). Yet ignoring a relevant area of study provides no viable solution. It means in reality falling back on the conventional wisdom within one's own field, whose assumptions may nonetheless prove open to question in the face of the newly emerging areas. And, of course, it may be for precisely this reason that so much inertia and resistance can be found to these newer approaches. But current behavioral science theory will not be salvaged by ignoring unpleasant developments elsewhere; and, more to the point, it will fail to account for human thought and action if its basic premises turn out to be untenable in the light of neurobiological research.

The second choice, that of presuming that future findings will prove compatible with existing theory, is unacceptable for largely the same reason. It is much more likely that current behavioral science theory (as well as much of sociobiological theory) will have to be qualified in the light of an emerging neuroscience rather than vice versa.

The third possibility will in fact eventually materialize, that is, the new knowledge of the brain will be incorporated within the social sciences. Even now the new dramatic developments in the biochemistry of the brain and in psychopharmacology demand attention by social scientists, if only on the level of public policy. But a systematic incorporation of neurobiological knowledge into the social sciences is still premature. The former area is still too new, too changing, and too unfinished to be ready for such a synthesis, while the latter is still too unreceptive and unprepared to receive it. And even if both fields were ready, the undertaking would be an imposing one. If Sinsheimer (1971) is correct in saying that the human brain may not be fully capable of comprehending itself, then the task of fully comprehending the relationship between the brain and behavior may be even more difficult. "The human brain and the behavior correlated with its development, structure, and function present a problem," Schmitt (1967, p. 2) writes, that is "the most complex to which man can address himself through the use of scientific methods." Since Sinsheimer, moreover, is implicitly assuming the best scientific minds available to address this problem, those of us with lesser faculties find ourselves in an even more untenable position. The rampant specialization within the sciences may well have a genetic and neurological basis. How many Renaissance minds are or will be available to understand the task of synthesizing neurobiology and social science?

Until the time becomes ripe and such Renaissance minds appear, the continued technical compartmentalization of the various fields appears largely inevitable. Yet social scientists cannot completely ignore the ramifications of new developments within neuroscience. In such cases, just as the political scientist must now accept the political impact of economic factors while leaving their explication to the economist, the social scientist must likewise, I believe, fully accept the social, political, and policy implications of biochemical and physiological factors while leaving their technical explication to others.

This expedient, however necessary or desirable, nonetheless fails to address the larger question of the validity of the basic assumptions with respect to neurobiological theory currently made within the social sciences. As Hirsch (1963, p. 136) has maintained, the "models and assumptions" of the behavioral sciences "must be consistent with the knowledge that is burgeoning at other levels," including "developments in fields which may once have seemed remote from behavior but which clearly are not." Only the last alternative, I believe, can render social science theory in general consistent with neurobiological research and theory.

The final position involves relying on the more general and accessible writings of the acknowledged leaders in neurobiology and relating these to the social sciences. I find this choice the most congenial. It is realistic—it can readily be done; it is somewhat tentative—as it should be, given the inevitability of later developments; and it is sensible in scientific terms—it means correcting positions even now untenable in the light of current neurobiological theory. I believe that these conclusions apply across the board to behaviorist psychology, to sociobiological theory, and to behavioral science. Accordingly, the next chapter will examine the basic thought of leading contemporary neuroscientists.

NOTE

1. This critique applies as well to their later work, *Promethean Fire* (Lumsden and Wilson [1983]; for a critical review see Elliott White [1984]). In April 1983, E. O. Wilson made a speaking tour of Philadelphia in connection with the publication of this work, and I raised with him the following question: Granting that both genetic and cultural factors influence our behavior, might not the mind, as an emergent product of the brain—as such leading neuroscientists as R. W. Sperry hold—operate as a causal force in its own right and therefore exert downward causation of its own in addition to genetic and cultural forces? In reply, Wilson indeed noted that there were "tough-minded" brain scientists who endow the mind

with causal potency. His own predilection nonetheless was to view the mind as basically explained by genetic, physiological, and cultural influences. But he also granted that this question was an important and an open one, and he went on to concede that should the emergent position come to be scientifically supported, then he would be constrained to follow it.

2

The Organism

The influence of mind on the doings of life makes mind an effective contribution to life. We can seize then how it is that mind counts and has counted. That it has been evolved seems to assure us that it has counted. How it has counted would seem to be that the finite mind has influenced its individual's 'doing'. Lloyd Morgan, the biologist, urged that 'the primary aim, object and purpose of consciousness is control.' Dame Nature seems to have taken the like view.
—Charles S. Sherrington (1963)

A systematic synthesis of the subject matters of the neurosciences and the social sciences seems premature; but at the same time the central role of the brain in accounting for human thought and action, as increasingly suggested by neurobiology, dictates that this role be fully acknowledged within the social sciences. At the least, then, the basic premises underlying the social sciences should be made consistent with current neurobiological theory.

Accordingly, I will proceed in this chapter to examine the basic thought of the following leading neuroscientists: R. W. Sperry, Sir John Eccles (with an inevitable reference to his part-time collaborator on the subject, philosopher of science Sir Karl Popper), and Karl Pribram. Others whose contributions to contemporary neurobiological theory require mention here include Ragnar Granit, Paul MacLean, and the late Wilder Penfield. Two points should be made at the outset: First, like any group of individuals, scientists or otherwise, these brain scientists do not necessarily agree with one another on any given point—but as it happens, they do form a consensus on precisely the question most relevant to the study of human society. Second, the fact that they do form a consensus does not assure the validity of their position or that, even if they should prove to be basically right, that that position will not be importantly qualified or refined in future years. Nonetheless, those of us who are not authorities in the area should ask ourselves one simple question: Upon what basis should we substitute our judgment for theirs, especially if our judgment is dictated by the demands of a theoretical approach whose own validity is at best problematic?

The one question that is most critical for the social sciences and that neurobiology has increasingly come to answer with one voice is the question of whether consciousness is an emergent property of the brain with an independent causal status of its own. As Sperry (1980, p. 206; cf. Sperry 1986) puts it, "Of all the questions one can ask about conscious experience, there is none for which the answer has more profound and far-ranging implications than the question of whether or not consciousness is causal. The alternative answers lead to basically different paradigms for science, philosophy and culture in general."

Sperry (1980, p. 197) sets forth an ascendant neurobiological position as follows: "Current concepts of the mind-brain relation involve a direct break with long-established materialist and behaviorist doctrine that has dominated neuroscience for many decades. Instead of renouncing or ignoring consciousness, the new interpretation gives full recognition to the primacy of inner conscious awareness as a causal reality." After describing this "new interpretation," I will proceed in the second half of this chapter to explore the larger implications of this approach for the question of human freedom. I will indicate that as opposed to either a free will or determinist position, one of self-determination is called for.

For Sperry in particular and for contemporary neurobiology in general, the first explicit and prominent formulation of a mentalist position dates back only to 1965. At that time Sperry (1965, pp. 77–78) openly admitted that his position aligned him "with the 0.1 percent or so mentalist minority in a stand that admittedly also goes well beyond the facts." In reiterating these views little more than a decade later, Sperry (1976a, p. 52) was to note that they had "come to acquire considerable acceptance and support."

A MENTALIST MODEL OF THE BRAIN

Inasmuch as Sperry's 1965 presentation of a mentalist position now appears to be its most classic, influential, and graphic formulation, let us briefly review it. Sperry (1965, p. 78) here contended that "mind and consciousness are dynamic, emergent (pattern or configurational) properties of the living brain in action" and that "these emergent properties in the brain have causal potency." The decisive issue, according to Sperry (1965, p. 79), is "who pushes whom around in the population of causal forces that occupy the cranium":

There exists within the cranium a whole world of diverse causal forces; what is more, these are forces within forces, as in no other cubic half-foot of universe that we know. . . . The various atomic elements are "molecule bound"—that is, they are hauled and pushed around by the larger spatial and configurational forces of their encompassing molecules.

Similarly, the molecules of the brain are themselves pretty well bound up and ordered around by their respective cells and tissues. . . . Even the brain cells, however . . . do not have very much to say about when they are going to fire their messages, for example, or in what time pattern they will fire them. The firing orders for the day come from a higher command.

Sperry (1965, p. 80) concludes that " . . . if one keeps climbing upward in the chain of command within the brain, one finds at the very top those over-all organizational forces and dynamic properties of the large patterns of cerebral excitation that are correlated with mental states or psychic activity." Consequently, Sperry (1965, p. 82) asserts that "near the apex of this command system in the brain" are ideas: "In the brain model proposed here, the causal potency of an idea, or an ideal, becomes just as real as that of a molecule, a cell, or a nerve impulse. Ideas cause ideas and help evolve new ideas."

Sperry's brain model is therefore both mentalist and emergent—mentalist in acknowledging the reality and causal force of mental states, and emergent in identifying such mental states with the higher level functioning of the brain as a whole and nonreducible to its component parts (or indeed to any other lower level or external set of factors). Sperry (1980, p. 195) also insists that his position is monist with respect to the mind-brain question; that is, unlike the dualist, who affirms the existence of mental states independently of the physical brain, the monist "says 'no' to an independent existence of conscious mind apart from the functioning brain." For my own part, I do not intend to address the complex philosophical mind-brain question. What Hilgard (1980, p. 15) writes regarding "psychologists and physiologists" fully applies to social scientists, namely, that they "have to be modest in the face of this problem that has battled the best philosophical minds for centuries." Moreover, it will likely remain an open question for the foreseeable future, for as Penfield (who himself gravitated from a monist to a dualist position) wrote (1975, p. 117), "a final conclusion . . . is not likely to come before the youngest reader" of his 1975 work dies.

For the social sciences the critical point remains, as Sperry's mentalist views indicate, that a monist as well as a dualist position may insist upon the independent and causal role of human consciousness. In this key respect, Sperry *does* support the mentalist position of Sir John Eccles, if not its dualism. Sperry (1980, p. 197) reports that after he sent his "new mind-brain 'answer' " (the 1965 formulation just reviewed here) to Eccles, "who previously had expressed little, if any, active interest in the holist-reductionist issues," Sperry "was delighted" to see that by 1968 Eccles "had clearly joined our ranks as an ardent antireductionist denouncing 'the materialistic, mechanistic, behavioristic, and cybernetic concepts of man' '': "Reversing his earlier stand on the uselessness of consciousness for a full account of brain function, Eccles has since lent his support to the new logic for the causal influence of mind over neural activity. On these points I believe we have remained in good general agreement." Now it is precisely this "general agreement" that is of the most direct interest to the social sciences.

In turning attention to the work of Eccles, then, I am more interested in its mentalism than its dualism. Eccles (1976a, p. 117), like Sperry, stresses the emergent character of the human brain: "The subtlety and the immense complexity of the patterns written in space and time by this system are beyond any levels of investigation by physics or physiology at the present time . . . and perhaps for a long time to come." A materialistic mechanism is for Eccles (1976b, p. 160) out of the question; it is "an ancient superstition according to which man is the victim of iron determinism as defined by nineteenth-century physics."

Eccles (in Popper and Eccles 1977, p. 373; cf. Eccles, lecture 2, 1980) summarizes his basic position as follows:

Its central component is that primacy is given to the self-conscious mind. It is proposed that the self-conscious mind is actively engaged in searching for brain events that are of its present interest, the operation of attention, but it also is the integrating agent, building the units of conscious experience from all the diversity of the brain events. Even more importantly it is given the role of actively modifying the brain events according to its interest or desire, and the scanning operation by which it searches can be envisaged as having an active role in selection. Sperry [1969] has made a similar proposal.

The influence of the ideas of Sir Karl Popper, perhaps the foremost living philosopher of science, on Eccles has been pronounced (see Eccles 1973, 1979, 1989). For that matter, Sperry (1980, p. 200) acknowledges that his own concepts of "mental phenomena as causal determinants in brain processing are extended and enriched particularly in the upper linguistic and epistemological levels by the insights of Popper." Popper, in turn, has been strongly influenced by Eccles. Both collaborated to produce the provocative *The Self and Its Brain*, published in 1977. The meaning of the title of the book is explicated by Popper (Popper and Eccles 1977, p. 120) as follows: "I intend here to suggest that the brain is owned by the self; rather than the other way around. The self is almost always active. . . . The mind is, as Plato said, the pilot. It is not, as David Hume and William James suggested, the sum total, or the bundle, or the stream of its experiences. This suggests passivity."

Popper's critical reference to the passivity of the Jamesian stream of consciousness (also repeated later [Popper and Eccles 1977, p. 517] in dialogue with Eccles) finds support in the work of Wilder Penfield, who also emphasized the active and independent role of human thought. Penfield's treatment of the Jamesian stream of consciousness metaphor as "a river, forever flowing through a man's conscious waking hours" made the point. For Penfield (1975, p. 49) found this metaphor misleading: "A river of water cannot be altered by the man on the bank. But thought and reason and curiosity do cause the stream of consciousness to alter its course and even change its content completely. The biological stream that is hidden away in each of us follows the command of the observer on the bank. . . . It is the mind (not the brain) that watches and at the same time directs."

For Eccles, the Popperian formulation of the three worlds has been particularly influential and has been integrated within his theoretical approach; that is, there exist interactive relationships between the first world of physical objects, including the human brain; world 2 of states of consciousness; and world 3 of objective knowledge, including all of man-made culture. Indeed, this Popperian framework transforms Eccles's dualism into a "trialist-interactionist hypothesis" (Popper and Eccles 1977, p. 360). Thus there is interaction between the physical brain and subjective states of mind (worlds 1 and 2); between worlds 1 and 3, as when man-made knowledge takes physical form in, for example, books and buildings; and between worlds 2 and 3: "When the self-conscious mind is engaged in creative thinking on problems or ideas, there would seem to be a direct interaction of world 2 and world 3" (Popper and Eccles 1977, p. 360).

It bears repeating and emphasizing here that it is possible for neuroscientists to recognize the reality and causal significance of conscious awareness quite aside from the adoption of a monist, dualist, or even trialist position. Let us explore this possibility further in the monistic though nonmaterialistic position of Karl Pribram. In Pribram's view (1971, 1976a), it is "the basic function of the brain to generate the roles"—the languages that "are the key to the structure of consciousness"—by which information (which "is not the property of any single event, but the property of the relation between them") is communicated (1976a, p. 301).

Pribram acknowledges that "many biologists, including Sir Charles Sherrington, Wilder Penfield, Sir John Eccles, and Roger Sperry, are dissatisfied with this sort of explanation because they cannot as yet visualize a brain mechanism that readily transforms nerve impulses into subjective experience" (1976a, p. 300). But Pribram also makes clear that his own reductive approach does not deny the reality or efficacy of conscious awareness:

. . . we ask not how brain and consciousness interact, but how the organization of interaction of basic brain elements differs in the states characterized by automatisms and those characterized by consciousness. As noted already, this form of reduction is not a pernicious reductionism that denies reality to consciousness or "explains" all the manifestations of consciousness in brain terms. Conscious awareness is a realization as real as is brain. In understanding the origins of the organization of consciousness we employ reductive procedures leading to the structure of brain, but in understanding the organization of brain we employ procedures that are equally reductive and which lead to the structure of awareness. And who is to say that one of these reductions is more fundamental than the others? Or who would claim that these reductions provide the total panorama of the realities we call "consciousness" and "brain"? (Pribram 1976a, p. 303)

Pribram's position, in short, is not, as he explicitly states, "a reductive materialism." Rather, it affirms that scientists are "ready (and capable) to defend spirit as data" while anticipating that "the days of the cold-hearted, hard-headed technocrat appear to be numbered" (1976a, p. 312). As Maxwell (1976, p. 342), responding to Pribram's position, concluded: "Pribram has emphasized that consciousness is important, practically important, even. It is very heartening to hear this coming from a tough-minded scientist. He comes to grips with Ryle's 'ghost in the machine' and comes to the conclusion that ghosts (of the kind that rile Ryle) really exist, and that they too are important."

Whether taking a monist or dualist stance, then, a brain scientist may acknowledge the existence and importance of what Popper (Popper and Eccles 1977, pp. 19, 20), following D. T. Campbell (1974, p. 180), terms "downward" causation within a hierarchical structure, and hence the "macro structure may, as a whole," act upon its parts. Thus Sperry (1976a, p. 54) writes that "as is the role for part-whole relations, a mutual interaction between the neural and the mental effects is recognized. The brain physiology determines the mental effects, as is generally agreed, but also the neurophysiology is at the same time reciprocally controlled by the higher organizational properties of the enveloping mental operations."

This model of mind-brain interaction is endorsed by Sperry's distinguished student, Gazzaniga (1988, p. 14): "A thought can change brain chemistry, just as a physical event in the brain can change a thought." Thus Gazzaniga (1988, p. 10) accepts his mentor's view that "the emergent properties of brain, the operating rules of the system we call mind, can push information around in such a fashion that the actual functioning of the nerves can be influenced by what the mind does." Gazzaniga (1985) does somewhat qualify Sperry's wholistic brain model by stressing what he sees as the "modular" organization of the brain; that is, that the brain is made up of discrete subsystems that can function in parallel; but their emphasis does not, as we have seen, repudiate Sperry's "macrodeterminism."

If the whole may influence its parts, and accordingly the higher level the lower level, the newly emerging and emergent features of an organism and its brain may also influence its older and more established properties.

Consequently, as was noted in chapter 1, MacLean (1973, p. 58) has indicated that the higher and later levels of the brain may influence the earlier and more primitive—and not simply the reverse. And we may also recall that Schmitt (introduction to Mountcastle [1978, p. 2]) summarized the theoretical formulation of Mountcastle (1978) by stating that "although phylogenetically older parts of the brain can play a significant role, the key to any fruitful theories of higher brain function must be the unique structure and properties of the cerebral cortex."

The acknowledgment of downward causation permits, in short, an independent causal role to human consciousness. As Granit (1977, p. 213) puts it: "No sooner is one—with Sherrington—prepared to look on consciousness as an emergent novelty, the greatest of nature's innumerable inventions, than it becomes imperative to regard it as a cause on a par with other causative agents, moving muscles in speech and laughter, in posture and precision grip, and thus also moving the world."

DETERMINISM, SELF-DETERMINISM, AND FREE WILL

From the perspective of contemporary neurobiological theory, then, the empty organism becomes the autonomous individual. The biological body becomes directed by the brain whose highest level property is that of self-awareness. Such a perspective places a central importance on what Popper (1979) calls world 2, the subjective realm of conscious awareness. For here reside the conscious purposes of man, the ideas that Granit suggests may move the world.

If this perspective is indeed valid, the implications for the social sciences can hardly be more profound. But precisely for this reason and for the reason that these views represent, as Sperry (1981, p. 7) observed, "a direct break with long-established materialist and behaviorist doctrine," the social scientist will naturally be hesitant before adopting them. Even their scientific legitimacy as theory may be questioned: Are we not, one might ask, returning Ryle's "ghost" to the "machine," Skinner's "homunculus" to the body, and E. O. Wilson's "vitalistic metaphysics" to the study of the organism?

Does not an emergent mentalist model of the brain mean sacrificing a scientific determinism to a vitalist free will position? I can no more hope to resolve the eternal free will/determinism conundrum than the mind-brain question; yet I must clarify the claims of a mentalist neuroscience in order to support their scientific status. In so doing, I will attempt to show that a mentalist approach conforms neither to a determinist nor to a free will position; rather "human decision-making" is, in Sperry's words (1980, p. 200), "self-determinant." In order to contrast a self-determinist position with one that is either free will or deterministic, let us briefly consider these positions in the light of the alternative models of the brain discussed in chapter 1.

Determinism

Each of the materialist reductionist models referred to—environmentalist, genetic, and physicochemical—is clearly deterministic in character. There is an ironclad causal order within the universe that includes, therefore, the human brain. Yet insofar as a reductionist model necessarily views human consciousness as epiphenomenal—by reducing it to a prior internal or external set of factors—it might better be termed *predeterministic* in nature. Our mental states are invariably the effect of prior causal conditions; they are never causal agents in their own right. Both sociobiology and behaviorism, we are reminded by Searle (1978, p. 168), "are deterministic about all human behavior in the strict sense that they see each action as caused by antecedently sufficient conditions, and they both reject any mentalism in the explanation of behavior." Such a predeterminism further suggests that to confer causal status on mental states is somehow unscientific.

These observations are clearly evident in Skinner's behaviorism. "A scientific analysis of behavior," writes Skinner (1971, p. 196), "dispossesses autonomous man and turns the control he has been said to exert over to the environment." What becomes abolished in the process, Skinner (1971, p. 191) asserts, is not biological man, but rather "autonomous man—the inner man, the homunculus, the possessing demon, the man defended by the literatures of freedom and dignity." Therefore "consciousness is a social product. It is not only *not* the special field of autonomous man, it is not within range of a solitary man" (Skinner 1971, p. 83; emphasis in the original). It follows, furthermore, that "abstract thinking is the product of a particular kind of environment, not of a cognitive faculty" (Skinner 1971, p. 180).

Behaviorism can hardly be more antimentalist. To assume the causal potency of ideas and feelings is to assume, for Skinner (1971, pp. 11, 12; emphasis in the original), the inner man, who becomes "a *center* from which behavior emanates. . . . We say that he is autonomous—and, so far as a science of behavior is concerned, that means miraculous." And thus inherently unscientific.

A sociobiological approach shares, as we have seen, a similar materialist, antimentalist stance. It is perhaps more tentative, as well as, of course, biologically based. But it, too, ultimately rejects a mentalist causation. Thus philosophers, like everyone else, Wilson (1978, p. 6) writes, find themselves influenced by "the deep emotional center of the brain . . . located just beneath the 'thinking' portion of the cerebral cortex" that has in turn "been programmed to a substantial degree by natural selection over thousands of generations."

As Barash (1979, p. 203) points out, sociobiological theory posits ultimate control in the genes: "Our genes have given us more freedom than those of any other living thing. Yet, just as the horse is still finally controlled by a rider, it is unlikely that we are completely free. Unbridled freedom would probably be a bad strategy; genes have undergone rigorous selection for millions of years, and those that have survived have done so because they had something of value to impart. We are free, it is true, but free only to maximize our fitness and that of our silent genetic riders." Similarly, Alexander (1979, p. 132) contends that the choices that we subjectively feel to be freely made are in fact dictated by inclusive fitness calculations performed internally within the central nervous system.

Wilson (1978, pp. 73–77) concedes that in practice the scientific understanding of the causal basis for our activity may be too difficult to attain, but that "theoretically it can be specified." Such a specification means for Wilson (1978, p. 78), however, the rejection of a "too complex neurological model" that "can lead back to a vitalistic metaphysics, in which properties are postulated that cannot be translated into neurons, circuits, or any other physical units."

A similar antimentalist, antivitalist orientation, finally, can be clearly associated with a physicochemical model of the brain. "Thus, for any event we might describe as an action and explain by citing its reason," Rosenberg (1980, p. 199) writes, a mechanistic, physicalist, "neurophysiological theory provides the resources to construct an alternative explanation of the same movement that makes no mention of desires, goals, motives, intentions, beliefs, hopes, or fears." As with both behaviorism and sociobiology, vitalism is an unscientific taboo. In his most reductionist phase, Crick (1966, pp. 98–99) noted that with respect to the nervous system "vitalistic ideas not only are commonplace among educated laymen, but are held by several of the leading workers in this field." With increased knowledge, Crick continues, "and when the study of fast computing machines have advanced even further," then "vitalistic ideas about the brain will grow to look as peculiar as vitalistic ideas now seem in molecular biology." Later, as a student of the brain, Crick (1979, p. 224) warns against "the fallacy of the homunculus," observing that "we certainly have the illusion of the homunculus: the self"—which may indeed "reflect some aspects of the overall control of the brain," the nature of which "we have not yet discovered."

Before proceeding, we might note a tendency, whether deliberate or otherwise, to associate a mentalist position that can be scientifically supported with unscientific straw men, e.g., "homunculus" or "vitalistic metaphysics." In effect mentalism is denied scientific status through the technique of guilt by association, when in fact even the association is problematic. As we shall now see, a mentalist approach may be distinguished from both free will and determinist positions.

Free Will

For a definition of free will, I turn to the *Dictionary of Psychology* (Warren 1934, p. 110): "The theory that the course of thought and volition is or may be directed by the individual himself, regardless of external influences and (according to some) regardless of internal or mental constitution."

This admittedly extreme definition implies an unfettered, unembodied "will." It implies, in short, a vitalistic dualism. Such a position has been associated most prominently with French philosopher Henri Bergson and his *élan vital*, and Hans Driesch, the German biologist of the early twentieth century. The latter, for example, postulates the "psychoid" as "the acting something which we have discovered not to be a machine" and which is "the elemental agent" that directs the body (Driesch 1908, p. 86). Indeed, Driesch asserts that "the common opinion about life-phenomena, which of course is neither analytical nor theoretical in any cause," which "claims that I can move my body by my will and that every living being has a so-called soul by which it can do the same" (and which Beckner [1968] refers to as "naive vitalism"), can "now be said to have been transferred from a non-analytical and non-theoretical to an analytical and theoretical sphere, and to have been proved and psychologically justified in this sphere."

This vitalistic dualism is of course rejected out of hand by a materialistic monism. Ryle (1949, p. 63) sees the idea of "will" or of "volitions" as "just an inevitable extension of the myth of the ghost in the machine. It assumes that there are mental states and processes enjoying one sort of existence, and bodily states and processes enjoying another." In any event, Mayr (1982, p. 52) observes that "for biologists vitalism has been a dead issue for more than fifty years."

Self-Determinism

Among contemporary neurobiologists this position is most clearly identified with Sperry. For Sperry (1980, p. 200) human decision making is "self-determinant": "Everyone normally wants to have control over what he does and to determine his own choices in accordance with his own wishes. This is exactly the kind of control our mind-brain model provides." Sperry's position avoids a vitalistic dualism at the same time that it rejects a materialistic behaviorism (cf. Dennett's "self-control," 1984, ch. 3). It claims, moreover (Sperry 1965, p. 92), to be "an objective, explanatory model of brain function," in other words, at least capable of empirical refutation. If indeed supported in future neurobiological research and theory, it would, as Sperry (1980, p. 205) has stated, have "profound and far-ranging implications."

Most directly, it introduces teleological explanations as accounts of our actions. Our consciousness, Etkin (1981, p. 67; cf. Etkin 1985) points out, permits us "to transcend the limitations of time and space," and consequently we are oriented toward an anticipated or desired future:

As a consequence human behavior is not a completely mechanistic system for it is, in part, determined by what is interpreted as part of the future. Our behavior is thus to some extent goal directed, capable of changing direction to keep in line toward something that does not yet exist. In short, we are teleologically directed . . . in so far as our anticipated goals existing only as images in our consciousness guide our present behavior.

It might, however, clarify matters to refer, following Mayr (1976), to teleonomic rather than to teleological explanation. The latter term may, consistent with certain vitalist views, refer to the actions of an organism that are determined by ultimate purposes that transcend the organism and inhere, for example, in the evolutionary process or in the universe more generally. Teleonomy refers, on the other hand, to goals that originate in and are limited to the organism itself and that nonetheless may motivate it.

Teleonomy and self-determinism are for humans closely associated. Both deny a wholly mechanistic approach to life; yet both accept the full application of physical and physiological laws. As Sperry (1980, pp. 201, 202) writes, "The mental forces do not violate, disturb or interfere in neuronal activity but they do supervene." An analogy that Sperry employs involves the transmission of television programs. The laws of electronics fully govern this transmission, yet they do not help "to explain why Mary struck John on channel 4, or what caused the building to collapse on 2, or the laughter on 7." Rather there are "higher order, supervening, program variables" that do "control at each instant, and determine the space-time course of the electron flow patterns to the screen and throughout the set—just as a train of thought controls the patterns of impulse firing in the brain."

Ironically, Ryle (1949, pp. 76–81), in a subchapter entitled "The Bogy of Mechanism," provides an analogous illustration. Ryle asserts that laws of nature "govern everything that happens" and yet "do not ordain anything that happens"; Ryle then asks us to imagine a chess match at which a scientifically trained spectator sits. This spectator, who is unfamiliar with chess, is permitted to look at the chessboard only in the intervals between moves; he therefore does not see the players actually make the moves. After a period of time he will nonetheless discern certain regularities, which of course correspond to the lawful movements of the various pieces. He may be tempted to conclude that "every move that you make is governed by unbreakable rules" and therefore "the whole course of what you tragically dub your 'game' is remorselessly preordained." The players, Ryle says, "laugh and explain to him that though every move is governed, not one of them is ordained by the rules: 'There is plenty of room for us to display cleverness and stupidity and to exercise deliberation and choice' " (cf. Clark 1980).

Sperry's self-determinism also fully acknowledges our obedience to all physical and physiological principles, while insisting upon our capacity to utilize these principles in exercising our choices. This does not mean, one must add, that our choices are free from genetic, physiological, and environmental constraints—although neither are they fully determined by them.

Sperry's monism is to be distinguished, as has been pointed out, from the dualism of Popper and Eccles. I will not elaborate this distinction here; rather I will only note that Popper and Eccles clearly intend their position to be taken as a scientifically defensible theory (I should also add for the record that I do not myself endorse—*pace* Schubert [1989, p. 323]—the "dualist theories of brain function such as the one that Popper coauthored with Eccles").

Sperry also distinguishes his causal determinism from Popper's indeterminism. "In contrast to Popper," Sperry (1980, p. 200) writes, "I hold that every time the elements of creation, whether atoms or concepts, are put together in the same way under the same conditions, that the same new properties would emerge and that the emergent process is, therefore, causal and deterministic. To this extent and in this sense it may also be said to be, in principle, predictable though generally, with few exceptions, it is not so in practice." In fact, however, Sperry and Popper may not be in basic disagreement on this point; for Popper (1979, p. 200; emphasis in the original) distinguishes between philosophical determinism—which would certainly embrace Sperry's position—and physical indeterminism. The thesis of the former states that "like effects have like causes" and that "every event has a cause." These principles, Popper asserts, are perfectly compatible with a physical indeterminism, which is "the doctrine that *not* all events in the physical world are predetermined with absolute precision, in all their infinitesimal details." Physical determinism, on the other hand, "asserts that the whole world with everything in it is a huge automaton and that we are nothing but little cogwheels, or at least sub-automatons, within it" (Popper 1979, p. 222). Clearly Sperry would join in rejecting such a predeterminism.

So would the neuroscientist Young (1987, pp. 209–11; emphasis in the original), who argues for a position that in effect corresponds to self-determinism. He sees individuals making creative choices between alternatives:

The essence of freedom is unforced choice by an individual according to his own character and needs. Each living organism is provided with a repertoire of possible programs of action. Everyone agrees that decisions between these as to what course to take are not made at random but on a basis of past and present information. The decision depends on factors acting *within* the organism by some form of balancing operation, about which neuroscience can yet say little in detail.... The outcome of this balancing operation depends entirely upon the characteristics of that individual, as built up over his own years of growth. A basic property of a human person is that most make choices. It is a responsibility imposed by human nature and individual characteristics, and the choice is free within the limitations of that nature and character. It depends not *only* on the laws that control the chemical events that are followed but on the particular pattern of combinations in which they occur; and *that* is the unique result of events of the immediate and remote past.

Inherent in the rejection of predeterminism is the acceptance of a good deal of unpredictability, especially in human affairs. Of course, in practice the predictability of a deterministic approach is itself at best problematic. As Wald (1965, p. 37) puts it:

When the time comes to make a decision, to exercise what we call free will, to choose—when that time comes, the self that exercises free will is, I think, that unique private self, that unique product of the unique composition, genetics and history, all to a degree unknown. At that moment no one can predict the outcome, neither an outsider nor the person making the decision, because no one has the requisite information. So I should say that the essence of free will is not a failure of determinism but a failure of predictability.

Now if one further stresses the independent causal status of conscious states of mind, then the element of unpredictability is greatly enhanced. Thus Winston Churchill described how the German ambassador to London was recalled for not having foreseen David Lloyd George's Mansion House speech of 1911, and commented: "How could he know what Mr. Lloyd George was going to do? Until a few hours before, his own colleagues did not know. Working with him in close association, I did not know. Until his mind was definitely made up, he did not know himself" (cited in the *Encyclopaedia Britannica*, 1962, S.V. "Free will").

Yet it is also the human mind that allows man some control over his destiny, what Popper (1979, p. 332) refers to as a "plastic control," in contradistinction to a "cast-iron control." Somewhat paradoxically, both the freedom and order in our lives are explained by this concept, for what we choose for our lives also imposes an order upon them that could not be imposed by any other source. If our behavior is at all determined, it is most directly and critically determined by our conscious purposes.

At this point we may summarize the preceding discussion by reference to Figure 2.1. The reluctance of social scientists to leave a predeterminist position is understandable when one realizes the degree of order, pre-

Figure 2.1
Three Positions on the Status of Human Freedom

	Position	Model of the Brain	View of Purpose
1.	Predeterminist	Reductionist materialist monist	Mechanistic
2.	Self-determinist	Emergent mentalist monist	Teleonomic
		or	
		Dualist/trialist interactionist	
3.	Free will	Vitalist dualist	Teleological

dictability, and control that one must relinquish as a direct result. But that order, predictability, and control exist only in theory. They do not appear to exist in fact. Therefore, to insist dogmatically on their existence is to employ, in Mayr's phrase (1972), a "retarding concept," that is, a concept that in fact gets in the way of a genuine understanding of ourselves. To deny "the existence or the importance of mental states," Seymour Kety (cited in Lieberman [1979, p. 332]; cf. Claxton 1988) has written, "merely because they are difficult to measure or because they cannot be directly observed in others is needlessly to restrict the field of the mental sciences and to curtail the opportunities for the discovery of new relationships. . . . Nature is an elusive quarry, and it is foolhardy to pursue her with one eye closed and one foot hobbled."

3

The End of the Empty Organism

The Coming of a Classical Social Science

> There is no scientific study more vital to man than the study of his own brain.
>
> —Francis Crick (1979)

> The brain is clearly so complex that the chances of being able to predict its behavior solely from a study of its parts is too remote to consider. The same complexity also warns us that the black-box approach or pure psychology will have to be lucky if it is not to bog down.
>
> —Francis Crick (1979)

> Limbic system research is not complete, by any means, and we must be careful about premature speculations. But what we know offers a more rational basis for informed judgment than anything suggested by the empty speculations of political scientists. Countries, nationalities, the names of leaders—all may change. But what remains is the common bond of the human brain.
>
> —Richard Restak (1979)

We human beings are endowed with brains whose most critical quality of self-conscious awareness is not reducible to any lower set of internal or external factors—whether molecular, genetic, physiological, or environmental in nature. Such has been the contention of the previous chapter, based upon examination of current neurobiological theory as represented by leading brain scientists. In subsequent chapters I hope to indicate that the role of the biological organism—the human individual—is literally integral to the evolutionary and historical process: It is a role that is active, critical, and indispensable. How, then, can the human organism—including most centrally its thoughts and purposes—be ignored in social science theory?

In this chapter I wish to anticipate and to outline the historical emergence of a social science that places the organism in a central role within its theory. This effort is admittedly speculative and not scientific; for as we shall see, neither the future source of organic evolution nor of human history is susceptible to scientific prediction. Yet, as I hope to suggest, trends are already in motion that are directly applicable to the future development of the social sciences. I argue that the most important area of research on the frontier of the life sciences today is that dealing with the higher nervous system and that this research will eventually, though it may take centuries, explicate what is now the "empty organism," explicitly or otherwise assumed within both the behavioral and the biobehavioral sciences. Using the language of Stent (1969), I further assert that the social sciences will come to incorporate the biological organism as a central concept into an emerging classical paradigm, well in advance of whatever may be the conclusive findings of a human neurobiology.

MOLECULAR GENETICS, HUMAN AND SOCIAL SCIENCE

There now seems to remain only one major frontier of biological inquiry for which reasonable molecular mechanisms still cannot be even imagined: the higher nervous system. Its fantastic attributes continue to pose as hopelessly difficult and intractably complex a problem as did the hereditary mechanism a generation ago. And the higher nervous system does, of course, present the most ancient and best-known paradoxes in the history of human thought: the relation of mind to matter. And so increasing numbers of veteran molecular geneticists are now turning toward the nervous system, in the hope that its study may soon enter a romantic phase, similar to that which attended the birth of molecular genetics. It seems likely that in the coming years students of the nervous system, rather than geneticists, will form the vanguard of biological research.

—Gunther Stent (1969)

The main thing science seems to be left with is the brain, and whether or not it is something more than a machine of vast and magnificent complexity.

It is a question that goes to the very center of man's being, so that fundamental changes in our view of the human brain cannot but have profound effects on our view of ourselves and the world.

—David H. Hubel (1979)

Psychology is essential. What the organism actually does we can learn only by observing it. Psychology alone, however, is likely to be

sterile. It must combine the study of behavior with parallel studies of
the inside of the brain.

—Francis Crick (1979)

When Gunther Stent, a little more than two decades ago, foresaw the
human brain as the last and greatest frontier in biology, he was accurately
reading the trends already in evidence to a perceptive observer. The last
ten years has more than amply supported Stent's forecast. As a result,
neurobiological assumptions currently made within the social sciences (see
chapter 1) are about to be put to the test by advances in neurobiology,
advances that are already being incorporated into the newly emerging field
of psychobiology, just as Crick (1979, p. 222) urges in the epigraph to this
section. What Crick advocates for psychology applies as well to the social
sciences. The attempt to understand the human organism through the
functioning of its brain can no longer be ignored by those of us concerned
with social and political life. As Restak (1979; p. 383; emphasis in the
original) insists, "How absurd that a person, in order to understand the
world, should study history, sociology, even psychology, yet possess no
information at all about the regulator of *all* human activities."

The direct and significant impact of the brain on our emotions, intellect, and behavior has already been conclusively demonstrated in a
number of ways. "It appears to be impossible," writes Boddy (1978,
p. 23), "to alter the physical state of the brain, whether by accident,
surgical intervention, administration of drugs or electrical stimulation
without in some way modifying conscious experience. If the brain is severely damaged the gross nature of the behavior that follows suggests
that consciousness has been severely degraded. Psychoactive drugs have
the capacity to modify the way in which we perceive the world, the way
we feel about what we perceive and our over-all mood, by altering the
physical state of the brain."

It is for good reason that any attempt to tamper with the brain, such as
through drugs or psychosurgery, is quite controversial and to be treated
very seriously. For an even more radical illustration, suppose, following
Popper (in Popper and Eccles 1977, p. 118), that successful brain trans-
plantations have become possible. Popper predicts (in principle) that such
brain transplants would likely transform the personality of the body oc-
cupied by the new brain. One might now ask oneself in this connection:
If I found myself near death as a result of some accident involving brain
damage and could only be saved through a brain transplant, would I accede
to one? Note that any possible fears or objections raised would reiterate
precisely the point that is being made here.

Yet the new developments in neurobiology seem foreign, confusing, and
altogether too complex for social scientists. If we cannot justifiably ignore
them, how can they best be related to our discipline? To attempt to answer

this question, I believe the best starting point is to gauge the present status of neurobiology in relationship to the present status of social science and then to anticipate, in the most general terms, the shape of the relationship to come.

I find most useful in this effort the model employed by Stent to describe the development of his field of specialty, molecular genetics, whose application he intends to be much more universal: "I will describe the history of my field in order to show that its rise and fall is but a paradigm of the history of creative activity in general" (1969, p. xi). Whether Stent's ambitious attempt does apply to all creative activity, I believe that it does have heuristic value for the life sciences and the social sciences.

Stent distinguishes four periods in describing the development of molecular genetics: (1) classic, (2) romantic, (3) dogmatic, and (4) academic. In the first or classic stage, genetics acquired an appropriate theoretical model with the formulation of the gene and its Mendelian principles. Classical genetics lasted from about 1865 through 1940. "During this entire development, however, the fundamental unit of genetics, the gene remained an abstract, formal concept, largely devoid of any physical content" (1969, p. 7).

The romantic period in molecular genetics dawned about 1940 with the beginning of a systematic inquiry into the biochemical nature of the gene. Genetics had now become a frontier science, attracting some of the best scientific brains from other fields, such as Max Delbrück, Linus Pauling, and Francis Crick. This frontier enterprise culminated in 1953 with the codiscovery of the structure of DNA by Crick and Watson.

The dogmatic period, lasting from 1953 to about 1963, takes its name from the central dogma of molecular genetics: that there is an irreversible one-way flow of information from DNA to RNA to protein. Through the application of this dogma, the functions of DNA became generally explicated.

The academic period, from 1963 through the present, is the direct outcome of the very success of the field: there are no longer fundamental questions to be answered. While molecular geneticists continue to do important work, "the would-be explorer of uncharted territory must direct his attention elsewhere" (Stent 1969, p. 66).

Thus, as Stent (1969, p. 71) has written (see epigraph for this section), neurobiology has become the major new frontier of the life sciences. This is not to say, of course, that other biological frontiers are thereby eliminated; developmental genetics especially has also become increasingly challenging: Certainly recombinant DNA research has become a frontier science—although it is in fact part of a developmental genetics that will come increasingly to embrace as well developmental neurobiology, that is, an understanding of the processes underlying the maturation of the brain. As Schmitt, Bird, and Bloom, the editors of the volume *Molecular Genetics:*

Neuroscience (1982), write in their preface: "Exciting developments that may profoundly alter the life sciences are currently being made in two fields: molecular genetics and neuroscience. The volume was developed to stimulate further advances by encouraging the application of current research concepts and techniques in molecular biology to the nervous system." Of course, the basic developmental processes underlying other physical organs will also be uncovered, so that the empty organism will be fully replaced by the real organism within the life sciences.

The implications may be striking. According to Schmitt, Bird, and Bloom (1982, preface), "If the explosive developments in molecular biology and neuroscience could be coupled, the impact on basic science and on clinical medicine would be very powerful indeed." Thus L. Thomas (1977, p. 167) finds it "difficult to imagine that this tremendous surge of new information will terminate with nothing more than an understanding of how normal cells and tissues, and organisms, function. I regard it as a certainty that there will be uncovered, at the same time, detailed information concerning the mechanisms of disease." At this point, then, it is appropriate to apply Stent's model to neurobiology, as well as to psychology and the social sciences.

By all accounts, neurobiology has just entered its romantic phase. The subtitle of *The Brain* (1979), by Restak, proclaims the field to be "The Last Frontier." Its frontier status is further signaled by the movement to its ranks of Crick from his earlier pioneer work in molecular genetics. Just what findings should emerge as neurobiology proceeds through its romantic stage—and, indeed, how many generations or even centuries will pass before these findings emerge—remain of course unknown. Hence, it makes no sense at all at this time, especially for a rank outsider, to hazard any speculations concerning the applicability of Stent's dogmatic and academic stages to neurobiology. Nonetheless, one obvious truth can be reiterated: In the immediate future and beyond, more and more will become known about the higher nervous system. Much of this knowledge will prove relevant to psychology and the social sciences, and consequently there will be less and less excuse in either area to continue to assume an empty organism.

Psychology, of course, is already becoming transformed into psychobiology and so entering its classic stage (Teylor 1975; Boddy 1978; Dimond 1978; R. Campbell 1988). As LeDoux and Hirst (1986, p. 21) observe: "In the last few years, there has been talk of a 'cognitive neuroscience', an interdisciplinary blend of researchers concerned with brain and cognition. Edited volumes have been compiled on the subject . . . meetings have been held . . . journals have appeared (*Brain and Cognition*), and the Society for Neurosciences now often includes a cognitive neuroscience section in the program of its annual meeting." Even psychiatry is moving in this direction as the biochemical basis for many mental disorders is being uncovered

(Kety 1979). While we cannot predict the speed and the substance of the findings to come, the general direction is once again unmistakable: psychology will become increasingly related to neurobiology. And the results of this convergence are bound to spill over to the social sciences, to which I now turn my attention.

In dealing with the social sciences, I will apply only the first two of Stent's stages; the classic and the romantic, since even more than in the case of neurobiology speculations concerning the final phases of the discipline are much too problematic. I would, however, add one stage to Stent's model, that of the "preclassic." This stage obviously precedes the classic; and while it may include some insights, it nonetheless encompasses a variety of differing and competing approaches, none of which validly and comprehensively describes the subject matter. Thus, none of the many ideas in circulation prior to Mendel adequately accounted for the principles of heredity. Until Mendel, then, molecular genetics was still in its preclassic era. Let me now apply these stages to the social sciences:

1. The preclassic stage. Social science is still, I fear, in its preclassic phase. The behavioral movement notwithstanding, there is no one central, generally accepted, and clearly valid theory (see Bell 1982). In political science this unsatisfying state of affairs is graphically described by Wahlke (1979) in his presidential address to the American Political Science Association (cf. Almond and Genco 1977; Almond 1990; Monroe et al. 1990). In sociology, Eckberg and Hill (1979, p. 925) call attention to the fact that "there are almost as many views of the paradigmatic status of sociology as there are sociologists attempting such analyses." Toulmin's characterization (1972) of the social sciences as "would-be disciplines" appears well founded.

2. The classic stage. Social science may well be on the verge of attaining its classic status. Indeed, Wilson would likely argue that such status was conferred with the publication of his *Sociobiology* (1975). No doubt classical social science will have an important biological dimension (how could it be otherwise?), and no doubt Wilson's work makes a significant contribution in this direction. But it may also be premature. As I have contended, sociobiology assumes an empty organism that is largely genetically preprogrammed. And as I have argued, these sociobiological assumptions concerning the brain are already untenable in the light of current neurobiological theory.

Sociobiology, moreover, accentuates its empty-organism approach by stressing ultimate as against proximate causation in social behavior. That is to say, the long-term evolutionary basis for behavior receives precedence over more immediate environmental and organismic conditions. Wilson's nomothetic vision of science, with its commitment to the discovery of universal patterns, further denigrates the rule of the individual. Wilson

(1975, p. 129), for example, desires "to establish that a single strong thread does indeed run from the conduct of termite colonies and turkey brotherhoods to the social behavior of man." How significant a role can human neurobiology as it relates to concrete individuals play in such a perspective?

Wilson, it seems to me, attempts to short-circuit the course of science that, as he himself (1977a) has implied, extends from particle physics through many body physics, chemistry, biochemistry, cell biology, and ultimately to social science. Wilson argues that each of these disciplines acts as an "antidiscipline," that is, in a "special adversary relation" to the discipline immediately adjacent to it at a higher level of organization. Thus, biochemistry attempts to reduce cell biology to its molecular level. Wilson (1977a, p. 127) argues that "the relevant branches of biology-neurobiology and sociobiology—are only now becoming mature enough to attain a juncture with the social sciences"; but he thereafter refers to biology generally as the antidiscipline for social science and ignores completely any explicit role for neurobiology. Indeed, even in a consideration of psychology and psychoanalysis, he discusses their relationship to biology only in the form of sociobiological theory, for example, "Psychoanalytic theory appears to be exceptionally compatible with sociobiological theory" (1977a, pp. 127–40).

Why, however, does Wilson leap from sociobiology to psychology without any mention of neurobiology? According to Wilson's logic, we ought to proceed from cell biology through neurobiology to psychology. Obviously, if one insists upon this logical progression, the role of sociobiology in accounting for human behavior becomes premature. Sociobiology will not, in any case, become generally applicable to human society until its basic assumptions become appropriately qualified in the light of an emerging neurobiology.

Social science, I believe, will enter its classic phase only when it directly treats the individual human organism in the light of the functioning of the higher nervous system. The nature of the biological organism, it seems to me, is the central concept to be formulated in classical social science. I do not deny that populations may have synergistic properties, as Corning (1983) correctly insists (and suggests may minimize the sociobiological emphasis on the role of the genes); but rather I question whether any theory concerning populations can validly be built upon the premise of an empty organism. Analogously, an emergent model of the brain must still be consistent with our understanding of the molecular structure of its component parts.

What is being advocated here is not simply the replacement of an S-R approach by one that reads Stimulus → organism → response. For the latter model implies an organism that is passive in its response to environmental stimuli; that is uniform within a species with respect to its response to a given stimulus; that is static in that no allowance is made for different responses by the same organism as it develops over time. In short, the *O* in the S-O-R approach might just as well stand for "empty organism."

If we grant that social science can no longer ignore neurobiology, how can we best address it? The view of Restak (1979, p. 54) in the epigraph to this chapter notwithstanding, I believe that it still is much too early to attempt a scientific integration of the two fields. That attempt in fact will herald the arrival of social science in its romantic phase. The most sensible course for social scientists to take is to adopt the basic theoretical model of the brain now propounded by leading neuroscientists that has been presented in chapter 2.

These emerging (and emergent) views of the brain clearly run counter to neurobiological assumptions held not only by behaviorist psychology and behavioral science, but also by much of biobehavioral science, including sociobiology. These views, nevertheless, must become incorporated into our study of human social and political life, and this incorporation will transform social science into its classic age. In the section to follow, I wish to hazard an outline of that transformation, but first I wish to risk an even more speculative comment on social science in its future romantic phase.

3. The romantic stage. The very first line of the introduction to *Brain and Human Behavior* (Kurczman and Eccles [1972] reads, "The disconcerting idea that dawned all of a sudden upon the Editors is that this volume may be premature by about 100 years." The synthesis of the areas covered by neurobiology, psychobiology, sociobiology, and social science clearly is generations into the future. Before social science is to enter its romantic era, neurobiology and psychobiology must first more or less reach the limits of their science. This will have happened not necessarily when everything is known, but rather when just about everything that can be known will have been known. There remains the question, after all, as Sinsheimer (1971; cf. Harth 1982) has stated, of whether the human brain can ever fully comprehend itself.

Only after the best scientific minds have exhausted these other fields will they then turn their attention to the social sciences. Thus, the status of social science as a scientific frontier is at the present quite remote. In the meantime, our task as social scientists is to formulate for our fields their classical theory. This task is surely challenge enough. And if much that we do now may be premature, we might also remember that brilliant insights may occur at any time and transcend any era: Tocqueville and Weber will surely be read a century hence. Before turning our attention to the larger implications of neurobiological theory, let me first present in summary form what I take to be the statuses of the fields of molecular genetics, neurobiology, psychobiology, and social science today from the point of view of Stent's stages of discipline development (see Figure 3.1).

Figure 3.1
Stent's Stages of Discipline Development

Stage		Field		
	Molecular			Social
	Genetics	Neurobiology	Psychobiology	Science
1. Preclassic	X	X	X	X
2. Classic	X	X	X	
3. Romantic	X	X		
4. Dogmatic	X			
5. Academic	X			

CLASSICAL SOCIAL SCIENCE

An emerging neurophysiology is bound to influence the social sciences; but should an emergent model of the human brain continue to find support, the social sciences will in the final analysis be most fundamentally affected by the need to recognize the autonomy and causal potency of human self-awareness. Social science will then ironically be forced to confront the living human being precisely as behavioral science has claimed to do. Easton (1953, pp. 201–2) wrote that the behavioral researcher "wishes to look at participants in the political system as individuals who have the emotions, prejudices, and predispositions of human beings as we know them in our daily lives. . . . Behavioral research . . . has therefore sought to elevate the actual human being to the center of attention." This behavioral focus on the individual has been reaffirmed by Eulau (1963, p. 44) and Wahlke (1979).

Behavioral science has failed in this mission. In systematically reviewing its findings, Berelson and Steiner (1964, pp. 666–67) remark on perhaps the most striking omission: "As one lives life or observes it around him (or within himself) or finds it in a work of art, he sees a richness that somehow has fallen through the present screen of the behavioral sciences." Behavioral science does not yet, they conclude, "see life steadily and see it whole."

But how could it be otherwise? Nothing in the past two decades fundamentally changes the verdict of Berelson and Steiner (cf. Bell 1982). Positivist scientific methodology applied to human life means that "all of its immediately subjective presentations to consciousness" is "deliberately circumvented or simply omitted" (Morison; cited in Berelson and Steiner 1964, p. 667). It no longer makes sense to permit a narrow methodology to dictate the subject matter of one's field. Not only has subjective consciousness been arbitrarily ruled out of bounds, but also all aspects of behavior that are not immediately assessable and measurable. The first order of business, however, remains to restore to our social and political theory the human consciousness that has never been absent from our lives.

According to Chomsky (cited in Baars 1986, pp. 346–47; cf. Claxton in Claxton 1988), the insistence that explanatory theories be "somehow reducible to observations" is "crazy" and "pernicious." Yet it is reflected in the very phrase "behavioral sciences":

It's a very curious phrase. I mean it's a bit like calling the natural sciences "meter-reading sciences". In fact a physicist's data often consists of things like meter readings, but nobody calls physics "meter-reading science". Similarly, the data of a psychologist is behavior, in a broad sense. But to call a field "behavioral science" is to say it's a science of behavior in the sense in which physics is a science of meter reading. Of course physics is not a science of meter reading; it's the study of matter and forces in the real world and so on. Psychology is similarly not the science of behavior, but the study of the mind, mental organization, and mental structure, which uses behavior as its data.

In this endeavor I foresee the development of a social science theory that grants both to itself and to its subject matter the following attributes: (1) purpose, (2) autonomy, (3) panclecticism, and (4) individuality. In the remainder of this chapter I will discuss these features.

THE PRIMACY OF PURPOSE: MEANING, ACTION, AND BEHAVIOR

Behavior can be recognized as act with all its intentional and intensional aspects. Not only is it once more respectable to investigate cognition, but a great deal is known about how the brain processes cognitions into perceptions and actions. If this last statement sounds Kantian, it is meant to. More and more evidence accrues to the effect that sensory input becomes processed into its component waveforms by a mechanism in which individual neurons or groups of neurons resonate to specific bandwidths or the frequency of the sensory input. Such resonators, as well as the transducer capacities of the sense organs, place limits on what is sensed as stimulus. At the same time other brain processes operate on the input, often preprocessing it prior to its organization into the mechanisms coordinate with conscious perception. Similar brain processes operate on mnemonic organizations in which are encoded waveforms generated by prior sensory input (the neural hologram). Such operations on the memory store are coordinate with

the cognitions. A hierarchy of these brain processes produce the syn-
tactic structures that program behavioral acts.

—Karl Pribram (1981)

On March 10, 1980, Jean S. Harris fatally shot Dr. Herman Tarnover.
Was it murder? Or an accident? The behavior itself is not in question but
its meaning is. On June 8, 1967, Israeli forces rocketed, napalmed, and
torpedoed the *Liberty*, an American intelligence ship operating fifteen
miles off the coast of Egypt, killing 34 American sailors and wounding
another 171. Was the attack intentional? Or was it accidental? The behavior
itself is not in doubt, but its significance is.

Social scientists cannot ignore overt behavior, but neither can they restrict themselves to its study alone. The distinctions that Reynolds (1976,
p. xv, emphasis in the original) draws, following Weber (1947), are especially instructive: "If we describe what people or animals do, without
inquiring into their subjective reasons for doing it, we are talking about
their *behavior*. If we study the subjective aspects of what they do, the
reasons and ideas underlying and guiding it, then we are concerned with
the world of *meaning*. If we concern ourselves both with what people are,
overtly and objectively, seen to do (or not do) and their reasons for so
doing (or not doing) which relate to the world of meaning and understanding, we then describe *action*."

We can only understand much of human behavior, as Midgley (1978,
ch. 5) makes clear, by understanding its underlying motivation. It is not a
matter of choosing between outer behavior and inner experience: "People
have insides as well as outsides; they are subjects as well as objects. And
the two aspects operate together. We need views on both to make sense
of either" (Midgley 1978, p. 112). Social science, then, takes as its field of
study human action; what is now behavioral science acquires a subjective
dimension: "The specific task of sociological analysis or of that of the other
sciences of action," writes Weber (1947, p. 941), is the "interpretation of
action in terms of its subjective meaning." The behavioral sciences, with
their emphasis upon overt behavior as the unit of study, became transformed following Parsons (1978, p. 5), who specifically includes psychology
within this classification, into the sciences of human actions.

The "explanatory understanding" of action (or *erklärendes verstehen*)
involves, for Weber (1947, pp. 94–96), a knowledge of the motive that the
actor attaches to his behavior. "Thus we understand the chopping of wood
in terms of motive in addition to direct observation if we know that the
woodchopper is working for a wage or is chopping a supply of firewood
for his own use or possibly is doing it for recreation." Such a *verstehen*
approach becomes an integral methodological component of a postbehav-
ioral or, in my language, following Stent (1969), a classical social science.

The hallmark of such an emerging science is the full recognition of the active role of the individual. Such a perspective places a central importance on what Popper (1979) calls "world 2," the subjective realm of conscious awareness. For here reside the conscious purposes of man, the ideas that Granit suggests may move the world. Of course, as Pribram (1976a, p. 301) points out, intentions and outcomes may differ: "The intentional universe is dispositional and may not even be realizable." Even unsuccessful attempts to influence the outer world are not without impact, however; and these too, therefore, have to be reckoned with by the social observer.

We conclude, then, by accepting the primacy of purpose in human action. Or as Popper (1979, p. 229) put it: "For obviously what we want is to understand how *intentions*, and *values*, can play a part in bringing about physical changes in the physical world." In chapter 6 I will further elaborate the implications of this position for the study of human affairs.

Sociobiologists and behaviorists alike offer compelling testimony to the primacy of conscious purpose. For how else do their prodigious professional labors become explicable except in the light of their goal of building a hard systematic science? And whence this goal? Was it innate? Was it imposed, *deus ex machina*, by the external world? Or even if biologically and historically influenced, was it not above all the product of purposively thinking individuals?

Indeed, add to the materialist theories of Wilson and Skinner the older ones of Marx and Freud. Each stresses a different aspect of the objective world—the gene (Wilson), environmental contingencies (Skinner), economic class (Marx), and the life and death instincts (Freud). Each prides itself as being hardheaded and scientific, even to the exclusion of all the others, and so presenting itself as the only sure guide to knowledge. Yet in the end they all remain theories, visions vastly influential in their time, in part because they legitimately point to factors that cannot be left out of the human equation, but also because of their concentrated conceptual focus—a focus ultimately responsible for their limitation and subsequent decline, yet initially one that served as a powerful attraction, a beacon, to those who would understand all by knowing only a part.

The Autonomy of Social Science

> I believe that the future will bring mutual illumination between two vigorous, independent disciplines—Darwinian Theory and cultural history . . . there will be no reduction of the human sciences to Darwinian theory and the research program of human sociobiology will fail.
> —Stephen Jay Gould (1980)

For both sociobiology and behaviorism, ethics and epistemology become reduced to evolutionary or environmentalist factors. They have no independent status or validity. Midgley (1978, pp. 169–72) suggests the inadequacy of the sociobiological position by focusing on mathematics. Wilson (1975, p. 3) asserts that both ethics and epistemology can be understood by reference to the evolution of the hypothalamus and limbic system that presumably largely controls their direction. Midgley (1978, p. 170, emphasis in the original) points out that "mathematics, too, is a branch of human thought, affecting conduct, for which no doubt certain specific parts of the brain are needed." Indeed, she writes, "the evidence for these parts being genetically determined is *much* plainer and *more* striking than it is in the case of ethics, since mathematics is a sphere particularly noted for infant prodigies." Therefore "these capacities have been subjected to natural selection, but it does not follow that the way to 'explain' mathematics and mathematicians—the 'fundamental' way is to dissect the brain and watch the neurones." Such a study may prove valuable in many ways, but not as a basis for the explanation of mathematics itself. Such an explanation "involves being able, first, to do mathematics; second, to grasp the standards that govern it . . . and, third, to relate mathematical standards to other standards of thought."

Mathematical thought must ultimately be understood on its own terms. Neither evolutionary nor environmentalist explanations are sufficient. Other systems of thought accordingly enjoy a comparative autonomy. In chapter 7 I will elaborate this position with reference to human ethics and politics.

An emergent model of the brain is entirely supportive of this position. We may recall and elaborate Sperry's contention (1965, pp. 82–83): "In the brain model proposed here, the causal potency of an idea, or an ideal, becomes just as real as that of a molecule, a cell, or a nerve impulse. Ideas cause ideas and help evolve new ideas. They interact with each other and with other mental forces in the same brain, in neighboring brains, and thanks to global communication, in far distant, foreign brains. And they also interact with the external surroundings to produce *in toto* a burstwise advance in evolution that is far beyond anything to hit the evolutionary scene yet."

An active and changing human consciousness may, then, dynamically interact with the sociocultural environment. In Popperian terms there exists an interaction between the subjective realm of world 2 and the world 3 of objective knowledge "when the self-conscious mind is engaged in creative thinking on problems or ideas" (Eccles, in Popper and Eccles 1977, p. 560). Such an interaction characterizes science itself; and while this fact hardly insures the validity of or the inevitability of progress associated with the scientific enterprise, it does mean the outcome of the interaction cannot be understood only by reference to the evolutionary or historical circumstances of which it is a part. Newtonian physics or Weberian sociology must ultimately be treated on their own terms and they are neither validated nor invalidated by reference to the particular biological, historical, and cultural circumstances in which they originate.

Of course evolutionary and historical considerations make a difference. A mathematical genius could more easily flourish in late Victorian England than in the early Middle Ages. If human brains have been selected for the limited and practical purpose of everyday survival, they may not be suitable for solving problems of cosmic dimensions, indeed for fully comprehending themselves (see Stent 1967). The apparent purpose of genes and life itself to survive by any opportunistic means suggests that any deception that proves self-serving will also survive. Accordingly, "claims made by scientists that they are interested in truth for its own sake," Hull (1980, p. 80) points out, "have to be viewed as sheer hypocrisy" (although Hull [1980, p. 80] in fact argues that "research scientists really do spend most of their time, money and energy doing what they claim to be doing—research"). Trivers (1981; p. 23, emphasis in the original) puts it starkly: "It is no good to imagine that scientists consist of a class of individuals created by society to pursue truth in a disinterested fashion. . . . Everything we know about evolution suggests that we have not been created to pursue truths in a *disinterested* fashion." We have here, it would seem, a case of the Cretan declaring all Cretans to be liars.

If, however, as Dawkins (1976, ch. 11) clearly concedes, genes alone cannot account for the rich diversity of human culture, and if cultural and indeed scientific evolution must be treated on their own terms, then science acquires a status of at least some independence if not objectivity. Dawkins asserts that a scientist (or artist) would rather be immortalized in human culture than within the human gene pool; and such an underlying motivation, while hardly insuring the permanence of any finished contribution, would surely promote the transcendent qualities that produce permanence. If most of us fail to leave our marks, it not likely through lack of trying (for, alas, "memes," like genes, do not always prove adaptive).

The attempt of sociobiological theorists such as Wilson and Trivers to immortalize themselves could not be more apparent. Their aims can hardly be more grandiose. Gould (1980b, p. 266) cites Trivers: "Sooner or later, political science, law, economics, psychology, psychiatry, and anthropology will all be branches of sociobiology." Nevertheless, Gould (1980b, pp. 266–67) clearly asserts the autonomy of human subjective awareness, including its role in providing meaning to our behavior (i.e., the behavioral "tears" make sense only in the light of the underlying experience):

I read Trivers' statement the day after I had sung in a full production of Berlioz' *Requiem*. And I remembered the visceral reaction I had experienced upon hearing the 4 brass choirs, finally amalgamated with the 10 tympani in the massive din preceding the great *Tuba mirum*—the spine tingling and the involuntary tears that almost prevented me from singing. I tried to analyze it in the terms of Wilson's conjecture—reduction of behavior to neurobiology on the one hand and sociobiology on the other. And I realized that this conjecture might apply to my experience. My reaction had been physiological and, as a good mechanist, I do not doubt that its neurological foundation can be ascertained. I will also not be surprised to learn that the reaction has something to do with adaptation (emotional overwhelming to cement group coherence in the face of danger, to tell a story). But I also realized that these explanations, however "true," could never capture anything of importance about the meaning of that experience.

Gould (1980b, p. 266) thus doubts that the social sciences will be "subsumed" as "mere epiphenomena of Darwinian processes." Rather he foresees evolutionary biology and the social sciences as continuing into the future to be "two vigorous, independent disciplines."

Indeed the dramatic rise of sociobiology has obscured the more quiet reemergence of a cognitive nonreductionistic orientation in recent years in a number of human sciences—often in reaction to a still current though receding behavioral approach but in effect rejecting sociobiological tenets as well. Thus Vernon Reynolds's *Biology of Human Action* (1976), published the year following Wilson's *Sociobiology*, was as a result overshadowed. It is, however, only one of many such works that have begun to affect the human sciences, ranging from psychology (Ornstein 1973; Dember 1974; Hilgard 1980; LeDoux and Hirst 1986; Claxton 1988; Hunt 1989); social psychology (Harré and Secord 1972); ethology (Griffin 1976, 1984); through anthropology (Dunn 1977; Freeman 1979); sociology (Berger and Luckmann 1967); and political science (E. White 1979; Schubert 1979, 1989; Losco 1981, 1982; Hines 1982; M. Edelman 1988).

With the exception of the psychological works, however, almost all of this cognitively oriented literature is surprisingly lacking in a foundation in contemporary neurobiological theory. (My colleagues Joseph Losco and Glendon Schubert are happy exceptions.) As I have ventured, social science theory will, following Stent (1967), emerge in its classical form when it replaces its present "empty organism" perspective with one that is fully consistent with prevailing neurobiological theory. Thus following Gould (1980b), I anticipate the continued autonomy both of the social sciences and of evolutionary biology. The latter in the form of sociobiological theory will continue to preempt the role of a reductionistic approach vis-à-vis the social sciences that hitherto has been reserved for behavioral science; and like the imperialistic forerunner of the latter, behaviorist psychology, it will probably require a good two generations to run its course and prove Gould's prediction (1980b, p. 267) that "the sociobiological vision of a reduction of one human sciences to biology via Darwinism and natural selection will fail."

Complementarity and Consciousness: A Plea for Panclecticism

> Having apprised himself of the latest research evidence from scientif-
> ically controlled studies involving hundreds of patients, the individual
> physician now confronts a single patient. At this moment the science
> and art of medicine converge. Clinical intuition and judgment, be-
> longing to the art of medicine, must make up for what we have not
> yet learned through research. But the physicians' artful acumen will
> always be needed, no matter how complete the scientific data, to apply
> general knowledge to a specific case and to enlist a fellow human in
> the actual task of healing.
> —Alan J. Gelenberg and Gerald L. Klerman (1978)

If self-awareness is a critical feature of the human species, then it must become a focus of study for social scientists. Nonetheless, if human self-awareness is not wholly reducible to molecular, genetic, physiological, or environmental determinants, neither is it wholly free of these influences. The full and explicit recognition of the critical role of the human organism within an emerging classical social science does not bring with it a single or simple unified view. Here is where Stent's model requires qualification: Although the concept of the gene has not proved to be entirely tractable and comprehensible in its scientific explication (cf. Brandon 1978), it is nonetheless a more restricted and coherent concept than one involving the human brain in all of its manifestations. The brain, more than the gene, will require multiple approaches for its understanding, taking into account all of the factors that may influence it—and leaving out none.

The two hemispheres of the brain themselves imply two different ways of knowing. While this distinction should not be overdrawn (cf. Sperry 1982), the left half of the brain tends to focus upon knowledge that is logical and analytical and the right half on understanding that is intuitive and holistic. Ornstein (1973, p. 10) draws the distinction sharply. He attempts to "document the existence in man of the major modes of consciousness: one is analytic, the other holistic. The first is analogous to the process of viewing the individual parts of the elephant, the second to viewing the whole animal. They are complementary; both have their functions."

Bohr (1963, p. 92) has explicitly argued that the study of life from both reductionistic and wholistic perspectives should be viewed as complementary rather than contradictory:

In biological research, references to features of wholeness and purposeful reactions of organisms are used together with the increasingly detailed information on structure and regulatory processes that has resulted in such great progress not least in medicine. We have here to do with a practical approach to a field where the means of expression used for the description of its various aspects refer to mutually exclusive conditions of observation. In this connection, it must be realized that the attitudes termed mechanistic and finalistic are not contradictory points of view, but rather exhibit a complementary relationship which is connected with our position as observers of nature. (cf. Weisskopf 1984)

Note, as Bohr points out, that the observer and his approaches themselves influence the determination of the final picture. In the social sciences, the works of Presthus (1964) and Allison (1971) illustrate the problem. Both writers compare and contrast competing approaches to problems in social science, in the one case, the elite-pluralist quarrel over the nature of the distribution of political power, and in the other, differing approaches to political decision making. In both cases, the writer assesses the strengths and weaknesses of the various approaches as he applies them to a particular place or event. But we also note that in making his judgments, the writer must transcend the particular theoretical and methodological limits of each approach. He here takes on the role of the artist, much as the doctor practices an art as well as a science in applying and also qualifying general principles to fit individual cases. Even the "hardest" science requires the intuitive judgment of the artist—and perhaps a reliance on the right sphere of the brain, in contrast to the more logical thought processes associated with the left hemisphere.

The indeterminist and unique nature of the subject matter especially necessitates this artistic as against exclusively scientific role for the social observer. This artistic component is to be welcomed rather than resisted, for it is what saves the scientific enterprise from being dashed against the rocks of a nihilistic relativism. For just as some doctors are more adept at the art of adjusting scientific generalization to fit the individual patient, some social observers may also better attain this art in their realm. A hierarchy of talent has a certain universal and objective basis (cf. E. White 1979).

Two artistic approaches to knowing that should receive increased attention by social scientists are ethological description and Weberian *verstehen*. "Every type of purely direct concrete description," writes Weber (1949, p. 107; emphasis in the original), "bears the mark of *artistic* portrayal." It was perhaps fortunate that Jane Goodall, before her classic study of chimpanzees in the wild (1971), had not been narrowly trained into a nomothetic science. Refusing to deal with her "subjects" in a wholly impersonal, objective manner, she described them with an eye to their individuality and their character. Somit (1976b, pp. 314–15; emphasis in the original; cf. Schubert 1989, p. 57), writing as a political scientist, observes that

ethology has made great advances by looking at actual animal behavior. This may stimulate political scientists, or at least some of them, to return to the study of *actual* political behavior. One of the great ironies of contemporary political science is that so many of those marching under the banner of "behavioralism" have turned their backs upon behavior itself. Political science may derive some profit by applying ethological concepts or even some ethological findings. It will profit to a much greater degree, I believe, if the ethological inspiration reorients us to the study of what people do, and how they behave, in day-to-day political life.

Verstehen represents an attempt, through empathy and imagination, to understand a unique historical event. As Weber (1949, p. 151) writes:

Das Kapital of Karl Marx shares the characteristic of being a "literary product" with those combinations of printers' ink and paper which appear weekly in the Brockhaus List—what makes it into an "historical" individual for us is, however, not its membership in the class of literary products but rather on the contrary, its thoroughly unique "intellectual content," which "we" find "set down" in it. In the same way the quality of a "political event" is shared by the hothouse political chatter of the philistine having his last drink at closing time with the complex of printed and written paper, sound waves, bodily movements or drill grounds, clever or also foolish thoughts in the heads of princes, diplomats, etc., which "we" synthesize into the individual conceptual structure of the "German Empire" because "we" turn to it with a certain "historical interest" which is thoroughly unique for us, and which is rooted in innumerable "values"—and not just political values either.

Artistic insights and descriptions may well have to be counterbalanced by more general and systematic accounts—and vice versa. Neither focus should be exclusive; both are complementary. What Isaiah Berlin (1976, p. xxv) says of history applies equally to the social sciences:

History needs whatever it can obtain from any source or method of empirical knowledge. As antiquarian research, archaeology, epigraphy, palaeography, philology have altered historical writing in previous centuries, so quantitative methods, the accumulation and use of statistical information to support economic, sociological, psychological, anthropological generalizations, have added to, and transformed our knowledge of the human past, and are doing so to an increasing extent. The use of chemical and biological techniques has added materially to the knowledge of the origins of men and the dating and identification of the monuments on which our knowledge is founded. Without reliable empirical evidence, the most richly imaginative efforts to recover the past must remain guesswork and breed fictions and romances. Nor is there any assignable limit to the influence upon historical studies of disciplines yet unborn. Nevertheless, without such inspired insights, the accumulated data remain dead: Baconian generalizations are not enough.

In the preceding remarks, I have implicitly called for an eclectic approach to the study of man. I wish now to expand the meaning of the term *eclectic*. To do so, I propose to coin a related word, *panclectic*. The etymology for the word *eclectic* derives from the Greek *ek* for "out" in combination with *legein*, "to choose"; therefore its meaning, as my Webster's *New World Dictionary* gives it, is "selecting from various systems, doctrines, or sources" and, second, "composed of material gathered from various sources, systems, etc." Unfortunately, for my purposes, the sense of the word can easily take on the notion of an activity that is haphazard, superficial, and dilettantish. One happens to come across various ideas and approaches and, with no clear or larger view or purpose in mind, simply collects those that have some special appeal. There is nothing inherently wrong in this activity, but neither is there anything especially desirable or necessary.

I propose a much stronger connotation by adding the Greek prefix *pan* or "all" to the Greek word for select. I have in mind here a double meaning for *all*. First, I wish to suggest that our selection of ideas and approaches applies in all cases, that is, we are always in the process of selecting material, at least if we view the selection process in the widest terms of including rejection as well as acceptance. Second, the ideas and approaches that we adopt can be applied to all manner of problem areas. With respect to human affairs, every conceivable activity is open to the application of differing approaches and, indeed, various combinations of approaches. Of course one particular approach or combination of approaches may be especially appropriate in a particular case, but the larger point is that no one approach can alone be applicable in every case.

The meaning and scope that I have in mind for *panclectic* may be made clearer by contrasting it with *monoclectic*, that is, selecting only one idea or approach and applying it across the board to a set of problems. A carpenter, for example, who took a strict monoclectic approach to his woodwork, might rely solely on a hammer to the exclusion of more suitable tools. Of course it is not difficult to see that a good carpenter must be panclectic, that is, he must choose from all his tools with a view to the problem that is before him—a hammer for pounding, a chisel for cutting, and so on; and his material may suggest further variations in the use of his tools—harder wood, for example, may call for a stronger cutting blade in a saw.

If a monoclectic carpenter is virtually a contradiction in terms, or at least is inevitably a poor carpenter, a monoclectic observer of human affairs must be similarly characterized. Kaplan (1964, p. 28) has formulated what he terms "the law of the instrument" as follows: "Give a small boy a hammer, and he will find that everything he encounters needs pounding." Substitute "social scientist" for "small boy" and a "new computer," "statistical technique," or "projective test" for "hammer" and you have the law applied to social science. Kaplan (1964, p. 29) goes on to say that "what is objectionable is not that some techniques are pushed to the utmost, but that others in consequence are denied the name of science." A narrowly monoclectic approach is necessarily self-limiting; and to insist otherwise is to insist on applying inappropriate techniques and perspectives indiscriminately over a wide range of problems.

A panclectic approach, on the other hand, takes full cognizance of all the ideas and approaches available for the study or treatment of a problem and is not a priori restricted to simply one (cf. Masters 1989, p. 1). In the treatment of mental disorders, for example, a patient who from early childhood has been severely depressed, regardless of circumstance, and whose family history includes frequent cases of depression may well benefit from a biopsychiatric approach involving drug therapy. The depression of a normally upbeat woman who has suddenly and unexpectedly been widowed following a happy marriage, on the other hand, might be more appropriately treated with another method. Other patients suffering from depression may benefit from a combination of methods. Indeed, Beck (1991, p. 370; emphasis in the original) argues that " 'mental disorders,' such as depression should not be considered as *either* psychomatic or biochemical, reactive or endogenous. A more sophisticated way of looking at such reactions is to recognize that psychological and biochemical phenomena are simply different sides of the same coin." The monoclectic therapist will apply only his specialty indiscriminately; the panclectic one will use to advantage the most appropriate technique available, ever mindful of the possible complementarity of various approaches. There has indeed grown up in recent years an "eclectic" school of psychotherapy, which attempts to treat its patients with what Lazarus (1990) refers to as a "multimodel" approach.

Of course few if any therapists or observers of human action can be conversant with every technique or approach available to them. But specialization and therefore monoclectic or perhaps "oligoclectic" therapists or observers are quite legitimate, provided only that they recognize their limitations and are ready to refer to or consult with others with differing approaches that may prove more appropriate. Still other practitioners or observers may be "polyclectic" and may therefore be better able to handle a wider range of problems. But because an underlying genetic and environmental variation also suggests a variation in abilities and aptitudes for employing various techniques and approaches, no individual, however polyclectic, is likely to be fully panclectic. It is for this reason that I refrain from employing the term *polyclectic* in contradistinction to *monoclectic* and as a substitute for *eclectic*, for I wish to uphold as the ideal a circumstance in which all the potentially applicable methods are considered available and none of which is excluded in advance for narrowly sectarian or ideological reasons. A merely polyclectic or, for that matter, eclectic approach fails fully to convey this broader vision.

A panclectic approach, in its broadest application to the study of human action, recognizes that the study involves both an art and a science and that therefore the insights and the techniques of both may be appropriate. The complementarity of both the artistic and the scientific, the mechanistic and the teleonomic, the etic and the emic (see chapter 6) are therefore recognized. A panclectic perspective indeed fully incorporates Niels Bohr's concept of complementarity (1963), which views such alternative approaches as the foregoing as being complementary rather than exclusive. To take a panclectic attitude is to insure that one will not pursue the study of human affairs, once again to quote Kety (in Lieberman 1979, p. 332), "with one eye closed and one foot hobbled." Every possible weapon in our intellectual arsenal will be at our disposal.

The Universality of Uniqueness

> Each person is an idiom unto himself, an apparent violation of the syntax of the species.
> —Gordon Allport (1955)

> Unlike any machine existing or contemplated, brains cannot be replaced, duplicated, or even reset. Thus not only is every brain unique, so is every moment in a brain's life.
> —Erich Harth (1982)

> Neurobiologists want to know how the brain works, but they don't say whose brain. Presumably when you have seen one brain you have seen them all. Given the extraordinary complexity of connections in a brain, it is at least conceivable, if not likely, that two people may organize their memories of the same event differently, or, God forbid, differently on different days of the week.
> —R. C. Lewontin (1983)

Thus far I have referred in my chapter titles to "the" organism, as if there exists just one standard type that is interchangeable one with another. Such typological references to organs or organisms are of course frequent within the literature, as witness, for example, the title of the *Scientific American* volume *The Brain* (1979). Yet some time ago Roger Williams (1956, p. 45) pointed out that the structures of our brains "often vary tremendously from one individual to another and that their structural likeness in general are under genetic control." More recently Childs (1975, p. 119) has referred to "the fermenting field" of investigating "the extent of polymorphism of brain tissue."

Sperry (1982, p. 1225) alludes to a "growing recognition of, and respect for the inherent individuality in the structure of human intellect": "The more we learn, the more we recognize the unique complexity of any one individual intellect and the stronger the conclusion becomes that the individuality inherent in our brain networks makes that of fingerprints or facial features gross and simple by comparison" (cf. Eccles 1980, p. 237; Harth 1982, pp. 96, 232; Eccles and Robinson 1984, pp. 42–44; Gazzaniga 1988, p. 63). Gerald Edelman (1987), a Nobel Prize–winning neuroscientist, has stressed the need to introduce population thinking to neurobiology, calling attention to "extraordinary variations" in brain structure and elaborating the immense evolutionary and behavioral significance of such variation.

A classical social science cannot evade the implications of such a basic individuality. In the succeeding chapters I hope to suggest the importance of human self-awareness within the context of human evolution and life, ever mindful of the vast, ever-present individual differences that characterize our species.

4

The Organism in the Evolutionary Process

The Implications for the Philosophy of Social Science

> We live, it appears, in a world of emergent evolution; of problems whose solutions if they are solved, beget new and deeper problems. Thus we live in a universe of emergent novelty; of novelty which, as a rule, is not completely reducible to any of the preceding stages.
> —Karl Popper (1979)

In this and the following chapter I wish to explore some of the larger theoretical implications that flow from an emergent mentalist model of the brain, especially when it is placed within a broader evolutionary framework. In this chapter I will focus on the ramifications stemming from such an evolutionary perspective for the philosophy of social science. In the next chapter I consider how the explicit incorporation of an emergent mentalist model of the brain within evolutionary theory qualifies evolutionary theory itself. As Hubel (1979, p. 3) has noted, "Fundamental changes in our view of the human brain cannot but have profound effects on our view of ourselves and the world."

Social scientists wed to established theoretical approaches are testimony to the power, indeed the tyranny, of ideas. John Maynard Keynes's well-known assertion comes to mind:

The ideas of economists and political philosophers, both when they are right and when they are wrong, are more powerful than is commonly understood. Indeed the world is ruled by little else. Practical men, who believe themselves to be quite exempt from our intellectual influence, are usually the slaves of some defunct economist. Madmen in authority, who hear voices in the air, are distilling their frenzy from some academic scribbler of a few years back. I am sure that the power of vested interests is vastly exaggerated compared to the gradual encroachment of ideas. (cited in Heilbroner 1953, pp. 4–5)

Sperry (1965, p. 92), who has enjoined us "never to underestimate the power of an ideal," would surely also hold to this statement.

Human populations vary and change over time. In addition, they are characterized by emergent properties on the individual and group levels, most dramatically in the case of the human brain. These biological attributes, finally, dictate an indeterminist social science, an indeterminism, one hastens to add, that does not imply an absence of causal order, but rather the inability to predict.

Such are the conclusions of this chapter. They are prompted by a consideration of the more philosophical writings of some of the most outstanding biologists of the past generation. The ideas carried in their work have in turn influenced some of the most prominent philosophers of science of the past several decades—Michael Polanyi, Michael Scriven, Stephen Toulmin, and Karl Popper. These philosophers of science are now coming to exert an increasing influence upon social scientists. In the field of political science, for example, Mulford Sibley (1967) and Samuel Hines (1979) prominently cite Polanyi, and Thomas Thorson (1970) is influenced greatly by Toulmin; John Gunnell (1969, 1979) is indebted among others to Scriven; finally Hines (1979) and especially Almond and Genco (1977) and Almond (1990) are intellectually beholden to Popper.

All of these philosophers of science, while differing importantly among themselves, are united in rejecting the positivist, determinist, and deductivist philosophy of science so prominently identified with the work of Carl Hempel (1968) and Ernest Nagel (1961). The latter hold that the main goal and practice of scientific inquiry (at least in its pure preprobabilistic form) revolves about the subsumption of individual empirical events within general laws; these laws in turn govern the occurrence of all such events in the present and the past (entailing "explanation") and imply, given the necessary specified initial conditions, their occurrence in the future (entailing "prediction"). The contextualist or indeterminist philosophers of science—who may even allow that such explanation and prediction are in principle possible—hold that in practice the complexity of the concrete interrelationships of events within the social universe imply constantly changing and significantly unique configurations that defy any general ordering—and that, moreover, to insist on abstracting such an ordering will result only in the formulation of static, or historical, and emptily formalistic laws (cf. Mandelbaum 1987).

What I intend here, however, is neither an explication nor a critique of these opposing philosophies of science—such are readily available elsewhere (see Toulmin 1977)—but rather a review of the concepts rooted in biology that have led away from a positivist and determinist scientific perspective. I will, however, focus, in conjunction with this review, on the relevant writings of Karl Popper. For Popper, in addition to being just possibly the most noted and influential philosopher of science now living, is also the only one to have conjoined within his approach both an explicitly evolutionary and a neurobiological perspective. I will also examine the relevant writings of the most distinguished evolutionists of the past generation. These evolutionists—who include Ernst Mayr, George Gaylord Simpson, the late Theodosius Dobzhansky, and the late C. H. Waddington—and their theoretical formulations will be conjoined with those of the neurobiologists that were considered in chapter 2.

Four main propositions flowing from this synthesis will be examined: first, that the subject matters of physics and of biology (and by direct inference, social science) differ significantly; second, that the validity of such key and uniquely biological concepts as genetic diversity severely limit the applicability to the social universe of nomothetic or "covering law" models of scientific inquiry; third, that these concepts call for, rather, an explicit appreciation of the full complexity of social order; fourth, that such an acknowledgment suggests the replacement of covering laws by "normic statements" within the social sciences; and, finally (in chapter 5), that the full acknowledgment of the unique character of human self-awareness within the context of organic evolution necessitates a theory of human evolution and of human history that focuses upon conscious purposes and what Popper (1979, p. 232) refers to as "plastic control" on the part of the human members of society.

It is understood once again, of course, that biologists, like members of any scientific discipline, vary and change over time. Accordingly, they do not all or equally endorse the above five statements. Nonetheless, there exists substantial support for these statements from among the leading lights within the discipline; and these in turn—while hardly exempt from error nor invariably saying something new or important—are in positions necessarily to command the respect and the attention of social scientists and, indeed, to state principles directly applicable to the human domain within the life sciences. As Simpson (1969, p. vii) writes, "Man himself can be considered as a biological phenomenon and a subject for biological study."

One further caveat is in order. While the main thrust of the biological ideas presented here is clearly supportive of Popper's basic philosophical position, there is not, of course, a complete identity of views. Thus in the section to follow, evolutionary biologists will be seen to object to a determinist physics that Popper (1979, p. 207) identifies by the metaphor of a "clock," that is, "regular, orderly, and highly predictable," and which he argues is in fact better represented by "clouds" or "physical systems which, like gases, are highly irregular, disorderly, and more or less unpredictable." Both Popper and the evolutionary biologists presented here are in basic agreement, however, that life is much more cloudlike than clocklike in character.

Popper's dualistic, indeed pluralistic, view of the mind-body question is also, as we have seen, problematic within contemporary neurobiological theory. But his emphasis upon the causal efficacy of conscious states of mind is strongly supported, as we have also seen, by leading neuroscientists. Let us now turn our attention to the first proposition to be presented: that the clocklike world of Newtonian physics is inapplicable to the life and social sciences.

CLOUDS, CLOCKS, AND BRAINS

> In place of the social scientists' favorite Myth of the Second Coming (of Newton), we should recognize the Reality of the Already-Arrived (Darwin); the paradigm of the explanatory but nonpredictive scientist.
> —Michael Scriven (1959)

The increasing prominence of the life sciences may or may not merit Simpson's boast (1969, p. vii) that biology, "and no longer mathematics, is now the queen of the sciences," but it surely supports the statement of his that immediately follows: "How far it has come, where it stands today, what it may become—these are of extraordinary importance among questions about man's intellectual attainments and prospects." Advances in biological knowledge have, in particular, reached the point where they can no longer be adequately understood by reference to scientific paradigms derived from the physical sciences. "Evolutionary organismal biology," states Simpson (1969, p. 7), "requires amplification of the philosophy of science to include its special features. It is not reducible to a philosophy taking account only of the physical, nonbiological aspects of the universe."

Three major contrasts can be made between the physical and the organic world. These contrasts, which are, according to Simpson (1969, p. 7), "of literally vital importance," relate to differences in (1) their level of complexity, (2) their degree of uniformity, and (3) the role of time as a significant factor.

Level of Complexity

"It is the complexity and the kind of structural and functional assembly in living organisms that differentiate them from non-living systems," writes Simpson (1969, p. 7), who goes on to postulate "a hierarchy of complexity" running from atoms, molecules, cells, and so on up through "the whole realm of the organic and its environments in space and time." At the pinnacle of this hierarchy of complexity, therefore, can be found precisely those biological phenomena with which social scientists are concerned, i.e., the set of evolving interactions characteristic of human life both on the individual and on the group levels.

This hierarchy of complexity is of consequence because it implies, in Mayr's words (1963a, p. 53), the "emergence of new qualities at higher levels of integration." Mayr states the principle involved as follows: "Where two entities are combined at a higher level of integration, not all the properties of the new entity are necessarily a logical or predictable consequence of the properties of the components." Thus the understanding of the functioning of an organism implies not only a knowledge of its constituent parts, but also a knowledge of their concrete and changing interrelationships, and these, in turn, as they develop, introduce reemergent qualities into the analysis that cannot be taken into account before they have in fact occurred. The very study of those interrelationships, moreover, may disturb them and thereby in effect extend the Heisenberg uncertainty principle into the organic realm.

The principle of emergence—or "compositionism," in Simpson's (1969, ch. 2) phrase (cf. also Corning's discussion of synergy [1983])—denies any reductionist position. In particular, it places scientists such as Simpson and Mayr—as well as Sperry and Eccles—at odds with those molecular biologists who see life as fully comprehensible in physical and chemical terms (cf. Monod 1972). More broadly, it leads them to question a philosophy of science grounded in the natural sciences. As Simpson (1969, p. 8) contends, "It is ridiculous to base a philosophy of science or a concept of scientific explanation wholly on the non-biological levels of the hierarchy and then to attempt to apply it to the biological levels without modification."

The rejection of a mechanistic or reductionistic position need not at all imply, as we have seen, an acceptance of vitalism, that is, the operation of transcendent, teleological forces in nature. It does, however, imply an explicit acceptance of the indeterminist, unpredictable character of much of life, especially as one ascends the hierarchy of complexity (cf. Popper 1979, ch. 6). For as the principle of emergence becomes increasingly applicable, an important consequence, Mayr (1963a, p. 53) points out, is that "it is certainly one of the major sources of indeterminacy in biology," by which Mayr means not "lack of cause" but "merely unpredictability."

A prime example of the properties of emergence is of course the human brain. Now the same complexity of the human brain that dictates its emergent character also dictates its inevitable uniqueness for each individual (cf. R. Williams 1956, p. 40; Popper and Eccles 1977, pp. 226–27), thus providing a further contrast with the organic world. The "very definition" of physical substance, Wald (1965, p. 23) points out, "is that all its molecules are identical." But this is not true, Wald continues, with living organisms: "There are no two living cells, and I would venture to say there never have been two living cells, that are or were identical." This "extraordinary individuality" of living organisms, Wald (1965, p. 24) further observes, derives from the fact of their being "enormously complex"—with that complexity being compounded by the fact "that living cells have not a static but dynamic composition." Out of the complexity of life, therefore, emerges the second major contrast with the physical world: its intrinsic individuality.

Degree of Uniformity

While the physical sciences "are for the most part typological," dealing generally "with objects and events as invariant types," Simpson (1969, pp. 8–9) asserts that the life sciences must deal with organisms of which "no two are likely ever to be exactly alike" and whose "diversity is not incidental." Mayr argues that this difference "brings home the enormous contrasts within science. In chemistry we deal with repeatable unit phenomena and with actions that, once correctly described, are known forever. In evolutionary biology we deal with unique phenomena, with intricate interactions and with balances of selection pressures—in short, with phenomena of such complexity that an exhaustive description is beyond our power" (1963, p. vi).

Such observations imply, as Dobzhansky (1967b, p. 41) puts it, a direct challenge to the assertion that has "gained acceptance by dint of frequent repetition" that "science is competent to deal only with what recurs, returns, repeats itself." Rather, Dobzhansky (1967b, p. 43) argues, "Individuality, uniqueness, is not outside the competence of science. It may, in fact it must, be understood scientifically." The science of genetics in particular, states Dobzhansky, takes individuality and its causes as its focal points; for a scientific understanding of the general principles of heredity for all sexually reproducing life includes the critical knowledge that those principles operate in such a manner as to insure genetic variability.

The necessity for the biologist to come to grips with the individuality intrinsic in his subject matter has necessitated a major paradigmatic shift within the life sciences. As Mayr (1970, p. 5) has written, "The replacement of typological thinking by population thinking is perhaps the greatest conceptual revolution that has taken place in biology." The distinction between these two modes of thinking is well set forth—though perhaps paradoxically a bit too typologically—by Mayr (1970, pp. 4–5) as follows:

The assumptions of population thinking are diametrically opposed to those of the typologist. The populationist stresses the uniqueness of everything in the organic world. What is true for the human species, that no two individuals are alike, is equally true for all other species of animals and plants. . . . All organisms and organic phenomena are composed of unique features and can be described collectively only in statistical terms. Individuals, or any kind of organic entities, form populations of which we can determine the arithmetic mean and the statistics of variation. Averages are merely statistical abstractions; only the individuals of which the populations are composed have reality. The ultimate conclusions of the population thinker and of the typologist are precisely the opposite. For the typologist, the type (eidos) is real and the variation an illusion, while for the populationist the type (average) is an abstraction and only the variation is real. No two ways of looking at nature could be more different.

Populational thinking within the life sciences has become necessary as the significance of genetic diversity has become increasingly recognized. According to Mayr (1989, p. 156), "the lavish production of diversity is the most important component of evolution." Its significance is tied to the integral part genetic variability plays in the evolutionary process. For if organisms and their component parts were, like molecules, uniform in composition, they presumably also would be unchanging. The fact that they are not makes it invalid, according to Mayr (1963a, p. 6), "to express the laws of evolution in terms of the laws of physics. Since every individual is unique, strict evolutionary reversibility is a logical impossibility." Hence the factor of time here emerges as critical.

Role of Time

If "the physical sciences are non-historical for the most part," Simpson (1969, p. 13) writes, "biology that is truly such, that is, a study of living things, inevitably and always has a historical factor, and the physical principles of repeatability, predictability, and parity of prediction and explanation do not apply to the historical aspects of biology." Mayr (1963a, p. 50) in effect concurs in pointing out that "one of the most important contributions to philosophy made by the evolutionary theory is that it has demonstrated the independence of explanation and prediction." In other words, explanation—or what Simpson (1964, pp. 146–47) terms "postdiction"—may well demonstrate patterns that have existed under certain necessary conditions in the past; but such patterns cannot be extrapolated into the future because the underlying conditions are ever changing. Since the occurrence of patterns involving life is necessarily time-bound, the ability to predict is also limited; for such ability would necessarily entail the capacity to foretell the development of unique configurations in the future.

Statistical probability provides no solution here. "Statistical methods are also widely used in modern biology," states Simpson (1969, p. 9), "but not as any counterplay against an immanent indeterminacy that makes their properties or activities inherently statistical." Statistics in biology are used rather to "measure individual variation and general organic diversity," and related phenomena. Prediction by statistical probability—which in any strict sense is no prediction at all—still assumes a certain qualitative uniformity of condition in the future, an assumption that is untenable.

These observations are made with respect to populations of individuals. Yet they are relevant on the individual level as well. If complex change through evolution characterizes populations (phylogeny), complex change through development characterizes the individual life cycle of the organism (ontogeny). Waddington (1968, p. 9) has contended that many definitions of the phenotype—often conventionally stated to be any observable characteristic of the individual—do not "clearly express the most fundamental and basic characteristic of phenotypes, namely that they change over time." Even the gene, conventionally viewed to be the one constant within life, may upon examination turn out to be highly variable (cf. Dillon 1983).

The phenomenon of life appears to involve, then, an interacting set of active individuals, all of whom are unique and complex, whose component organic parts also partake of these attributes and who change significantly on the individual and on the population level over time. Can a paradigm derived from classical physics do justice to this apparent reality?

GENETIC DIVERSITY, ENVIRONMENTAL DIVERSITY, AND EVOLUTION

> Where every individual has a unique genotype, the phenotypes are the results of unique systems of interactions, and the interactions of arrays of such individuals to each other and to the complex, ever changing environmental conditions are at each moment also unique. The relation of theory to observation will always be much looser than in the physical sciences. We may gain a much greater understanding of evolution than at present, but we will never predict its course in detail.
>
> —Sewall Wright (1978)

A paradigm stressing regularities over time simply cannot account for life that is unique, varied, and ever changing. It cannot come to grips with "the individuality that is so characteristic of the organic world"—an individuality in which, continues Mayr (1963, p. 52; cf. also 1978), "all individuals are unique, all stages in the life cycle are unique; all populations are unique; all species and higher categories are unique; all inter-individual contacts are unique; all natural associations of species are unique; and all evolutionary events are unique." Mayr goes on to assert that "where these statements are applicable to man, their validity is self-evident. However, they are equally valid for all sexually reproducing animals and plants." Social scientists implicitly employing a nomothetic model, however, have not regarded as self-evident the biological uniqueness both of man and of men; or if they have, they have done so only in a trivial sense. But if such individuality is in fact fundamental, then, as Mayr (1963a, p. 53; cf. Mayr

Figure 4.1
Patterns of Uniformity and Diversity in Genetic and Environmental Factors

		Environmental	
		Uniformity	Diversity
	Uniformity	1. (A clonal population)	2. An environmental determinism
Genetic			
	Diversity	3. A biological determinism	4. A populational interaction paradigm

1982, p. 846) maintains, "It is quite impossible to have for unique phenomena general laws like those that exist in classical mechanics."

The diverse and evolutionary character of living things is, in fact, of literally vital importance. "If organisms did not vary," Simpson (1969, p. 9) asserts, "and had not done so for some billions of years, they would not exist at all." For without variability, the adaptive mechanism of natural selection could not operate. Genetic diversity therefore plays an indispensable role in the evolutionary process.

Moreover, as Dobzhansky (1970a, pp. 232–33) maintains, "organic diversity is the adaptive response of living matter to the challenge of the diversity of environments." For if the world's environment were entirely homogeneous, Dobzhansky points out, "one kind of organism, a single genotype, would conceivably be sufficient to exploit the resources of the environment"—except that "even then the life activity of this single kind of organism might create an environmental heterogeneity and hence a place for another kind of organism." Both genetic *and* environmental diversity, in other words, are likely to characterize the evolutionary and, by implication, historical processes. Indeed, the interplay of diverse genetic and environmental factors over time *constitutes* those processes.

Yet few social scientists acknowledge the presence of both genetic and environmental diversity. At the least, however, the student of human society should make explicit—and then examine—his or her underlying premise in this key respect. Accordingly, let us visualize (in Figure 4.1 above) four possible patterns that the uniformity or diversity of genetic and environmental factors might take.

Let us briefly consider these four possible combinations in turn:

1. Genetic and environmental uniformity have never characterized human life. For if they had, the end product would necessarily have been an entirely uniform society—the existence of which cannot be empirically established.

Genetic and environmental uniformity have occurred (if at all) only in those limited instances in which identical twins have been reared together. Accordingly, to the extent that they may share a common heredity and environment, they tend to express concordant personality traits and behavior patterns (see Koch 1966). (But identical twins raised apart—or, for that matter, fraternal twins raised together—will invariably differ in some critical respects.)

The only theoretically possible occasion for the occurrence of both genetic and environmental uniformity involves a genetically engineered population of clones—and the very same clone at that, i.e., a world populated through the reproduction in mass form of a single individual (cf. Dobzhansky et al. 1977, p. 449; McKinnell 1978). In such a case genetic uniformity would tend to produce although by no means to insure an environmental uniformity as well, for each "individual" would be surrounded by identical "individuals."

Aside from the formidable (and fortunate?) practical difficulties attendant in bringing about this outcome, a population of clones would be novel from an evolutionary perspective: If each "individual" were alike and reproduced accordingly, evolution would cease. (Yet even here diversity would likely emerge, just as mutations produced variation at the outset of the evolutionary process when all reproduction was asexual and hence clonal.) In addition, all human history would end, for attitudes and behavior would not only prove uniform at any one time, but also would remain constant over time. At least to the extent that total uniformity in both the genetic and environmental realms was realized, then to that extent the historical process would come grinding to a halt. Such a static, uniform, and ahistorical social universe would indeed constitute a scientific dreamworld for the homothetically oriented behavioral scientist—if only it were to exist and if only the clone were not a creative genius!

A clonal population, however, might prove to be extremely dysfunctional from the standpoint of fulfilling societal needs, an observation that serves as commentary on the societal (and evolutionary) necessity and significance of human genetic variability. A final question that the reader might pose is, Would a clonal society have any politics? My own answer is that here only would a society bereft of a Madisonian faction or a Marxian classless utopia—or even Skinner's apolitical Walden II—have some chance of occurring. Nevertheless, what if, à la Freud, the clone were a rather disputatious character?

2. If human societies are not wholly monolithic, then at the very least either genetic or environmental diversity must exist. The environmental determinist, of course, accounts for any observed variability by variability in external circumstance. Thus Skinner (1953, p. 424) readily acknowledges the existence of an immense variation in society; but since individuals are basically alike at birth, such variation must be accounted for by environmental diversity. As Skinner himself allows, the social environment is

"probably never the same for two individuals." Such environmental diversity is likely to produce differential individual perceptions, which are the stuff of politics and which of course characterize existing societies. But they would also likely appear, along with political controversy, even in a utopian community of Skinner's own making. For if the argument is made that differing perceptions and hence political controversy might be eliminated by making the environment uniform, then, given the premise of genetic uniformity, one is in reality referring to the first set of conditions already discussed.

3. The biological determinist, on the other hand, may concede a genetic diversity; but inasmuch as heredity determines everything important, the environment becomes uniformly unimportant. Ardrey indeed takes this position. "A diversity of beings," he (1970, p. 38) writes, "encounters a singleness of being. . . . It is this singleness of the playing field that reveals the inequality of the players." Social and political differences are, from this view, literally inherent in human life; and, in Ardrey's case at least, so is a marked political inequality.

Sociobiological theory acknowledges both genetic and environmental diversity, but in practice is less interested in the latter. Human behavior that is common to all cultures is emphasized, at least by Edward Wilson (1978, p. 21):

our species is distinct from the Old World monkeys and apes in ways that can be explained only as a result of a unique set of human genes. Of course, that is a point quickly conceded by even the most ardent environmentalists. They are willing to agree with the great geneticist Theodosius Dobzhansky that "in a sense, human genes have surrendered their primacy in human evolution to an entirely new, nonbiological or superorganic agent, culture. However, it should not be forgotten that this agent is entirely dependent on the human genotype." But the matter is much deeper and more interesting than that. There are social traits occurring through all cultures which upon close examination are as diagnostic of mankind as are distinguishing characteristics of other animal species—as true to the human type, say, as wing tessellation is to a fritillary butterfly or a complicated spring melody to a wood thrush.

Cultural universals suggest, according to Wilson, a human biogrammar, a genetic determination of behavioral patterns unique to man. Such a focus tends, however, to obscure (though certainly not to deny) cultural variation, as well as intra- as against interspecific genetic diversity. Such a focus, quite aside from questions concerning its validity, has an undeniable appeal to those seeking a nomological science of man (cf. Rosenberg 1980, ch. 7).

The most graphic illustration of Wilson's emphasis upon the universal and the biological at the expense of the diverse and the cultural is in his discussion (Wilson 1978, p. 43) of "noncultural retardates," that is, individuals whose mental retardation is so severe as to deprive them of the capacity to participate in human culture. Human forms of communication—human language itself—are beyond them. Wilson proceeds to observe, however, that noncultural retardates "retain a large repertory of more 'instinctive' behavior, the individual actions of which are complex and recognizably mammalian," and these include the use of facial expressions and emotion-laden sounds in communication, the examination and manipulation of objects, masturbation, and territorial behavior.

Now just what, for Wilson, do such observations imply? Precisely because such behavior appears so "instinctive," it is therefore also taken to be most basic to our human nature; for Wilson (1978, p. 42) argues that human nature can be probed by examining "behavior that is both irrational and universal" and hence "less likely to be influenced by the frontal lobes and the other higher centers of the brain that serve as the headquarters of long-term rational thought." Hence the reference to noncultural retardation in order to illustrate such behavior and thereby presumably the most basic and universal attributes of human nature. It is instructive to contrast Wilson's typological approach here with Goldstein's examination of severely brain-damaged individuals in his *Human Nature in the Light of Psychopathology* (1963); for Goldstein discusses the pathological in order to distinguish it from and to emphasize the more normal. Thus the "essential capacity" of the human being for Goldstein (1963, p. 68) inheres in "the best functioning of the most complex part of the brain"—clearly a limiting perspective, and properly so—that excludes lower and more universal attributes that man in part shares with other animals.

4. One position alone escapes a reductionist, typological emphasis and recognizes fully the complexity of social order. An interaction approach grounded within a populational framework explicitly acknowledges the existence and interplay of both environmental and genetic diversity (cf. E. White 1972). Since human genetic variability is an established scientific fact, environmental variability follows as a necessary consequence. For if each individual is biologically unique, and if, as asserted in *Biology and the Future of Man* (Handler 1970, p. 431), the most important environment for each individual is that constituted by other individuals, it follows that the environment is likely to be unique for each individual.

We can now see in this context that the social scientist must come directly to grips with Mayr's assertion that each individual, each interindividual contact, and thus each evolutionary (and by implication, historical) event is unique. Yet we also face at this point the obvious and agonizing question: How can the scientist treat unique events and still remain a scientist? It is to this question that we now turn.

SIMPLE ORDER, COMPLEX ORDER, DISORDER

It is important therefore to ask, what Wordsworth found in nature that failed to receive expression in science. I ask this question in the interest of science itself, for one main position in these lectures is a protest against the idea that the abstractions of science are irreformable and unalterable. Now it is emphatically not the case that Wordsworth hands over inorganic matter to the mercy of science, and concentrates on the faith that in the living organism there is some element that science cannot analyze. Of course he recognized, what no one doubts, that in some sense living things are different from lifeless things. But that is not his main point. . . . He always grasps the whole of nature as involved in the tonality of the particular instance. . . . Wordsworth, to the height of genius, expresses the concrete facts of our apprehension, facts which are distorted in the scientific analysis. Is it not possible that the standardized concepts of science are only valid within narrow limitations, perhaps too narrow for science itself?

—Alfred North Whitehead (1968)

A sharp scientific distinction has frequently been drawn between the nomothetic and the idiographic, between the uniform and the unique. And yet we commonly come across the occurrence of both in our own everyday lives. "It is necessary to labour the point," observes Whitehead (1958, p. 5), "that in broad outline certain general states of nature recur, and that our very natures have adapted themselves to such repetitions." Yet Whitehead is quick to add: "But there is a complementary fact which is equally true and equally obvious—nothing ever really recurs in exact detail."

As against the simple order of the exclusively uniform and the extreme disorder of the exclusively unique, there may exist as middle ground the complex and complementary order of the uniform *and* the unique. And perhaps it is here, between the simple order of the classical physicist and the "absurd" disorder of the existentialist, that we may find the basis for a complex order suitable for the social and the life scientist.

On the one hand, science is not able to comprehend a world devoid of all order. As Wiener (1961, p. 50) makes clear, "For the existence of any science, it is necessary that there exist phenomena which do not stand isolated. In a world ruled by an irrational God subject to sudden whims, we should be forced to await each new catastrophe in a state of perplexed passiveness." On the other hand, science, especially any science dealing with life, cannot afford to emphasize the nomothetic to the exclusion of the idiographic. Dobzhansky (1970b, p. 31) states the case as follows:

Even in these days of science and technology triumphant, some philosophers have the audacity to restrict the competence of science to something less than the totality of the universe. Science, they claim, is concerned only with properties which 98 many things have in common, only with events which recur, return, or can be reconstructed and reproduced; the individual, the unique, the concrete happening, in other words the living reality of existence, is apprehended better by artistic, philosophic, and religious methods of cognition. Bergson has stated this view succinctly: "Science can work only on what is supposed to repeat itself." I confess a measure of sympathy for such an opinion, but must nevertheless point out that biology not only recognizes the absolute individuality and uniqueness of every person and every living being but in fact supplies evidence for a rational explanation of this uniqueness.

Dobzhansky's scientific synthesis of the uniform and the unique, or what I have called complex order, does not deny the principle of causality—that a precise set of conditions will always produce the same effect. It suggests only that within the organic world a precise set of conditions, once having occurred, will never again be duplicated. Hence each effect will be novel, thereby accounting for the reality, the irreversibility, and the unpredictability of the evolutionary and historical processes. The principle of causality remains intact, but an exclusively nomothetic approach succeeds only in grossly oversimplifying social and biological reality.

This analysis of complex order conforms fully with the philosophy of science of George Gaylord Simpson. Inasmuch as Simpson is one of the few eminent life scientists to have outlined explicitly such a philosophy, let us briefly review it. In keeping with the basic distinction between the nomothetic and the idiographic, Simpson (1964, pp. 119–21) distinguishes between the "immanent and the configurational." The immanent refers to those processes and principles within the material universe that are unchanging and therefore are "non-historical." The configurational, on the other hand, refers to "the actual state of the universe or of any part of it at a given time," which is, especially in its fullest description, "not immanent and is constantly changing."

The subject matter of the physical sciences is governed by immanent principles insofar as it involves processes that "are indefinitely repeatable, unchanging in character, and non-historical." But organismal biology—the study of living things—is necessarily historical and hence configurational in focus; and so clearly are the social sciences—and history itself, which Simpson defines as "configurational change through time, a sequence of real individual but interrelated events."

But are unique historical events then subject to scientific study? Simpson (1964, pp. 122–23) first of all understands science to be "an exploration of the natural universe that seeks natural, orderly relationships among observed phenomena and that is self-testing." He subsequently defines historical science "as the determination of configurational sequences, their explanation, and the testing of such sequences and explanations." Now, by "explanation," Simpson does *not* mean the subsumption of individual events under general laws. If "law" refers to "a relation or sequence of phenomena invariable under the same conditions," then Simpson maintains that the "search for historical laws" is "mistaken in principle." In addition to reasons stemming from the study of life already alluded to, Simpson explicitly enlists the work in the philosophy of science of Scriven (1959) (the validity of which Mayr [1963b, p. 50] also explicitly endorses). By contrast, Simpson (1964, p. 136) also explicitly rejects the view that he ascribes to Hempel and Oppenheim (1953) that scientific explanation and prediction are conjoined by overarching scientific laws.

To deny the parity of explanation and prediction is thus also to deny the applicability of explanation in the form of immanent law to historical science; rather, explanation must be construed in a qualitatively different sense for this context. Such a revised construct would not deny the relevance of the immanent to history, for "the uniformity of the immanent helps to explain the fact that history is not uniform" (the reader may here wish to reflect on Simpson's assertion). The construal of scientific explanation must be broadened in order to apply separately to the realms of the immanent and the configurational. Accordingly, Simpson (1964, pp. 133–34) endorses Nagel's distinction (1961) between (and justification for both) universal and contingent explanation.

This distinction roughly mirrors Simpson's own between predictive and postdictive explanation. While "non-historical science is mainly predictive," historical science is "mainly postdictive": "The most frequent operations in historical science are not based on the observation of causal sequences—events—but on the observation of results. From those results an attempt is made to infer previous causes. . . . Prediction is inferring results from causes. Historical science is largely involved with quite the opposite—inferring causes (of course including causal configurations) from results" (Simpson 1964, p. 146). Simpson (1964, pp. 146–47) suggests, then, that only the physical sciences remain in valid accord with a philosophy of science stressing the nomothetic and typological character of the universe. The historical sciences, on the other hand, must accept a complex order—in which "the individual event," in the words of Oppenheimer (1954, p. 93), "is a sort of intersection of many generalities, harmonizing them in one instance as they cannot be harmonized in general." To reintroduce the language and position of Simpson: Configurational events are the inevitable outcome of the operation and interaction of immanent principles.

We may now incorporate Simpson's analysis and its implications within our three categories of order as shown in Figure 4.2, below. Two observations on Simpson's dual classification of the sciences are now in order. The first registers an overdue caveat with regard to Simpson (and by implication most of the life scientists mentioned thus far). For in Simpson's classification, classical physics is taken as a prototype for the contemporary physical sciences, when in reality it may only represent the physics of a bygone century or, at best, a convenient method of dealing with a limited, albeit important, area of the physical world (though cf. Schubert 1983, pp. 109–10). Thus Popper (1979, p. 215; cf. Gleich 1989) argues, following Charles Sanders Pierce, that "all clocks are clouds, to some considerable

Figure 4.2
Categories of Order

	Simple Order	Complex Order	Disorder
Simpson's descriptive categories of scientific reality	The immanent only (the nomothetic)	The immanent and the configurational (both nomothetic idiographic)	The configurational only (the idiographic)
Philosophy of (or toward) science	Determinist	Indeterminist	(Existential)
Popper's scientific metaphor	Clocks	Clouds	
Mayr's modes of scientific thinking	Typological	Populational	
Simpson's classification of the sciences	The physical sciences (including classical physics)	The historical sciences (including the life sciences, the social sciences, and history)	

degree—even the most precise of clocks. This, I think, is a most important inversion of the mistaken determinist view that all clouds are clocks." As Cocconi (1970, p. 87) speculates concerning the likely future course of physics, "The immutable laws of physics could become as 'emphemeral' as those of organic life, immutable only for observations limited in space and time, and even more exotic, the evolution of these laws would depend on history, a history that has followed a path that, to a great extent, must have been determined by chance."

Ferris (1988, p. 334) observes that if a "supersymmetric—an invariant universe, preferably if characterized by an esthetically elegant proportioned universe" ever existed, it "belonged to the past": "the implication is that the universe began in a state of symmetrical perfection, from which it evolved into the less symmetrical universe we live in." Thus even physical invariance turns out, given sufficient time, to be historically variable!

We may therefore live in a Whiteheadian world in which the "process is itself the actuality" (Whitehead 1956, p. 173) and in which ultimately even physics becomes a historical science. Yet the question of degree may nonetheless loom critical here: The passage of a year counts for nil when it comes to an atom, but it becomes a fact of overwhelming import for any living cell. It is highly questionable, therefore, that sociobiology or any other organic life science may be treated, as Rosenberg (1980) would have it, as a hard deductivist science.

Our second observation follows: Considering the increasing efforts of some historians (cf. Landes and Tilly 1971) to tie themselves to the social sciences within a behavioral (and presumably from their viewpoint, properly scientific) framework, one might view with some sense of irony the contention by a leading life scientist that the biological and social sciences are in fact part of the historical realm and to be studied scientifically within a framework that implicitly rejects the basic tenets of the contemporary behavioral approach. For immanent laws operate so as to bring about genetic diversity and in consequence an environmental diversity as well. Thus, to paraphrase Whitehead, the diversity is itself the reality.

Therefore our social science must begin with the immanent, thereby avoiding the disorder of scientifically untenable propositions; but it must end with the configurational, thereby accepting the inescapable obligation to abide by the demonstrated implications of an ongoing scientific inquiry.

Scientific Explanation and Social Reality: Covering Laws, Normic Statements, and Contextual Analysis

Typological laws simply cannot do justice to social and historical reality. Before seeking their replacement with more appropriate generalizations, let us again consider why this is so. In an example of his covering law model of explanation, Hempel (1963, p. 358) sets forth the following propositions.

Agent A was in a situation of kind C.

A was a rational agent at the time.

Any rational agent, when in a situation of kind C, will invariably (or: with high probability) do X.

A did X.

Note the typological assumptions imposed upon any concrete circumstance: There is a uniform rationality applicable across the board to any situation; all rational agents partake of this rationality and hence can be expected to act accordingly in any given circumstance. Now take note of the likely reality of things: Just as some acts are more rational than others, so some actors are more rational than others. There is no doubt a wide array of rational capabilities and no two are perfectly alike. Were both JFK and LBJ rational agents? William Rogers and Henry Kissinger? U Thant and Kurt Waldheim? Nasser and Sadat? If any or all of these sets of individuals qualify as rational, were there still no significant individual differences dividing them that were of the utmost relevance for policy?

Now consider the environmental dimension. Granted that situations may be classified by kind (e.g., "kind C"), nevertheless to do so in an overly typological fashion is obviously to miss nuances that may in fact be more than nuances. The Munich Crisis of 1938 has been compared to crises in Korea, Vietnam, Eastern Europe, and the Middle East. Some comparability in some cases may even be allowed, but the differences must also not be overlooked.

Finally, Hempel concedes that the rational agent in question may perform a given act "with high probability" rather than "invariably." But this conclusion immediately acknowledges the possibility that all manner of other factors, impossible to predict, may interfere with the original expectation. Such indeed is the nature of the world of living creatures. We cannot even predict with certainty our own whereabouts a year from now. In any case, probabilistic statements are most appropriate for uniform populations of atoms or electrons; they are much less so for highly variable populations of unique, complex, and internally motivated individuals (cf. Elsasser 1966).

Indeed, even the physical world defies exact prediction. As Hawking (1988, p. 55) asserts: "The uncertainty principle had profound implications for the way in which we view the world . . . one certainly cannot predict future events exactly if one cannot even measure the present state of the universe precisely!" Is the alternative to such uncertainty, however, a chaotic, existential world in which nothing is related to anything else? The answer is no. As Grene (1969) writes:

All biological knowledge must deal with norms, which are standards and therefore universal, not particular. Even if the biologist deals, medically, with the patient here before him, or, genetically, for example, with populations statistically analyzed, he has still to have behind him the recognition of this or that kind of organism or this or that kind of behavior. He has to recognize, for example, the physiognomy of this illness, which must have some general, not just this particular, *Gestalt* if it is recognizable at all. He must know cv, redeye, or what you will from wild-type drosophila, or he could not begin to build up his populational statistics at all. Such biological universality, such normative judgments, must be integrated into the account of biological method at the start, not injected at the end.

The doctor cannot treat any patient without a general idea of the nature of varying diseases, their associated symptoms, causes, and likely cures. Yet the doctor must treat the specific patient and not just a general malady. To prescribe blindly an antibiotic to any patient, even if his condition in general appears to warrant it, is possibly to aggravate rather than to aid his condition.

Between absolute laws, on the one hand, and an existential void on the other, there is both the need and the justification for employing what Scriven calls "normic statements." Such statements apply to or account for behavior in the absence of "interference factors—or 'disturbing conditions,' or 'special circumstances.' " Of course to some degree "special circumstances" always qualify the normic statement, but they do not negate the fact that some factors do tend to relate to each other.

Normic statements, according to Scriven, are neither statistical nor conditional. They are not statistical, among other reasons, because they may refer to a multitude of circumstances, some of which may exist in an idealized form unsuitable for direct empirical analysis, and others of which may lack that uniformity of condition essential for statistical analysis. They are not conditional because, though in some limited instances all of the qualifying conditions that apply may be specified in their formulations, in general such a specification is not possible. Particularly within the social sciences an indefinite array of extenuating circumstances exists that defies explicit specification.

The possible application of normic statements to the social sciences may be illustrated by the example: "Intellectuals tend to be critics of society"— a proposition that has received substantial empirical support. Presumably such a relationship will continue to hold in the future. Yet it is also clear that not all intellectuals have been critics of society, that few have been critics on all occasions, and that indeed within some cultural traditions intellectuals have tended to uphold the ongoing order of things. What a normic statement does in such an instance is to alert us to the possible occurrence of the exceptional as well as of the more routine. It reminds us that in the final analysis one must examine the immediate context as well as the general one. Both particulars and universals, we recall, make up the complex order of human history, past and present.

In order to place these remarks in larger theoretical perspective, let us return to the four possible relationships that the uniformity or diversity of genetic and environmental factors might take, as has been indicated by Figure 4.1. Let us now add the connection between these relationships and possible formulation of either covering laws or normic statements.

What we observe above is that wherever we can assume some uniformity in either environmental or genetic factors, the covering law model of explanation is applicable, although whenever diversity still exists along either one of the two dimensions the application of the covering law must be made contingent upon the specification of the conditions that are in force.

Where there is uniformity in both the environmental and the genetic realms little such qualification of universal rules is presumably necessary. As we may recall, identical twins raised within the same family environment come about as close an any phenomenon in meeting this criterion in the real world. In such a case the observer can generalize safely from observations concerning one twin and the other.

Most of us are not identical twins, however, and family environments are never wholly uniform. Both genetic and environmental diversity are

Figure 4.3
Patterns of Uniformity and Diversity in Genetic and Environmental Factors as Related to Covering Laws and Normic Statements

		Environmental	
		Uniformity	Diversity
Genetic	Uniformity	1. Absolute covering laws	3. Conditional covering laws
	Diversity	2. Conditional covering laws	4a. Normic statements 4b. Absence of relationship

to some degree inevitable, therefore, and with such diversity the possibility of differing combinations of circumstances becomes staggering. Normic statements thereupon become applicable as general guides, at least whenever the diversity of circumstances along both the genetic and the environmental dimension does not become too pronounced, as suggested by the 4a position in Figure 4.3. When, on the other hand, the diversity of genetics and environmental circumstances does become too pronounced, no useful relationship can any longer be drawn; and indeed, the dissimilarities between events may well become more illuminating. Thus the social scientist must become sensitive to differences as well as to similarities, even between phenomena of like character. And as he does so, his study, like that of medicine, becomes an art as well as a science.

One additional point must be made explicit: The acknowledgment of the limits of scientific understanding may itself constitute a valuable contribution to scientific inquiry. It may allow us to marshall our energies by avoiding blind alleys. Mayr (1972), in referring to "retarding concepts," reminds us how the attempt to emulate Newtonian physics has, in the life sciences, made a full appreciation of the critical importance of organic diversity for the evolutionary process slow and difficult to grasp. More recently, Mayr (1982, p. 846) has reiterated that what he has called "physicalism" or the "erroneous search for laws" has had a "deleterious effect on development in biology"; he also alludes to "quintarianism" as "one of the many ill-founded endeavors to make biology 'scientific' by making it quantitative or by making it obey definite 'laws'." If Mayr is right, the further application to social science seems clear. Much wasted effort is already in evidence; but much more may yet be avoided.

Bronowski (1969, p. 41; emphasis in the original) calls attention to "how much positive knowledge we can derive from asking what we can not do," citing as an example the fact that "a great part of mechanics can be derived from the single assertion that *perpetual motion is impossible*." Accordingly, we should welcome rather than resist what Bronowski (1969) calls "Laws of the Impossible." If the social scientist realizes, for example, that the future will bring the novel, he at least knows enough not to expect the merely recurrent. (He who insists on reversing the premises had better brace himself for constant disappointment—no matter how much of a pragmatist he asserts himself to be.) "The maturity of a science," we may conclude with Price (1965, p. 106; cf. Mayr 1982, p. 846), is "measured not only by its power, but by its discrimination in knowing the limits of its power."

5

Organic Selection, Human Evolution, and Human History

Organisms do not experience environments passively: they create and define the environment in which they live. . . . There is a constant interplay of the organisms and the environment, so that although natural selection may be adapting the organism to a particular set of environmental circumstances, the evolution of the organisms itself changes those circumstances. Finally, organisms themselves determine which external factors will be part of their niche by their own activities.
—R. C. Lewontin (1978)

If there is one agency through which order may be imposed upon seeming chaos, it is the human mind. Of course, we may often impose more order than in fact exists. "The facts of the world in their sensible diversity are always before us," writes James (1948, p. 4), "but our theoretic need is that they should be conceived in a way that reduces their manifoldness to simplicity." Indeed, our brains may be too limited to recognize cases of extremely complex order. "Evolution selected this brain," Stent (1967, p. 114) asserts, "for the capacity to deal 'successfully' with superficial, everyday phenomena, but it was not selected for handling such deeper problems as the nature of matter or of the cosmos."

Nonetheless, a human self-awareness capable of recognizing its limitations clearly is not without some strengths. And just as clearly it must have evolved and have played an independently important role in the evolutionary process. In one of the last works on which Theodosius Dobzhansky collaborated, a text on evolution, the authors note (1977, p. 453) that "manuals of psychology and neurophysiology often do not even mention the slippery words 'mind' or 'self-awareness,' " but they immediately add,

"An evolutionist, however, is obligated to deal with them, since the emergence of mankind is incomprehensible without them." As Wright (1978, p. 396) asserts, "the central theme of human evolution during the last ten million years has undoubtedly been the increase in mental capacity, associated with a three-fold increase in brain size" (cf. Itzkoff 1983, 1985, 1987).

It is the evolution of the human mind that allows man some control over his destiny, what Popper (Popper and Eccles 1977, p. 232) refers to as a "plastic control," in contradistinction to a "cast-iron control." Somewhat paradoxically, both the freedom and the order in our lives are explained by this idea, for what we choose for our lives also imposes an order upon them that could not be imposed by any other source. If our behavior is at all determined, it is most directly and critically determined by our conscious purposes.

In this chapter, accordingly, I wish to indicate the role that conscious thought and purpose may play in the evolutionary process, especially in human evolution and, more directly, in the process of human history. At the same time, the important context involving genetic and environmental diversity is kept constantly in mind.

ORGANIC SELECTION AND EVOLUTION

If human thought can, as Granit (1977) suggests, change the world, it can affect the evolutionary process. Indeed, artificial selection is man-made evolution, as Charles Darwin pointed out in the very beginning of his *Origin of Species*. Animals are bred to conform as closely as possible to human preferences. Human choices also have affected and will continue to affect the evolution of species other than those we have bred for domestication. Whether we attempt to save an "endangered species," eliminate a harmful virus (from our perspective), or indeed create one inadvertently in the process of genetic engineering or deliberately for the purposes of germ warfare, we shall, in any event, influence the course of evolution more generally.

In these and other ways man also influences his own evolution. Boehm (1978, p. 40; cf. Sternberg 1985, pp. 50–52; 1988, pp. 14–18, 266–69) argues that efforts to control the environment

have made and will continue to make a difference in the adaptive possibilities, probabilities, and outcomes experienced by human groups. These outcomes are of enormous potential importance to the continued adaptive effectiveness of human populations—and possibly to their very survival. Because such purposive and insightful interference into the very process of evolution anticipates and bypasses natural selection processes, this kind of conscious, intentional meddling will be termed "rational preselection."

Whether infrahuman species may also be characterized by "rational pre-selection" is a question that I leave open. Boehm (1978, pp. 269–71), for one, does extend this capacity to other species, specifically describing Hamadryas baboons in these terms. Griffin (1976; 1984) (see also Schubert 1979) has directed "the question of animal awareness" and its possibility to the level of the honeybee. Human rational decision making is of course not at issue here, but noninstinctive, flexible, and purposive actions are.

Insofar as purposive behavior and lower forms of consciousness characterize animal life in general, the evolutionary process no longer should be viewed as blindly mechanistic. According to Dobzhansky (1970b, p. 93), "Even lower animals move where the conditions please them." Each organism, Darlington (1969, p. 269) states, "seeks and selects the place, the environment, that fits it best." Of course not all attempts at self-selection will succeed; and indeed Mayr (1970, p. 358) points out that every "major shift of habitat is an evolutionary experiment" and that "most of them are failures." Nevertheless, Mayr (1982, p. 573) asserts that "what is most remarkable is that bearers of different gene arrangements not only have a different fitness in different niches but also the behavioral capacity to search out the right niche."

To the extent that organisms can take actions that are not wholly constrained by genetic or environmental factors, they may affect their own chances for survival. The organism then plays an active role in the evolutionary process and ceases to be wholly the passive product of blind external forces commonly summarized under the term *natural selection*. Rather what has been termed *organic selection* emerges—whose roots may be traced back to Morgan (1896), Baldwin (1897), and Hardy (1965). This theory asserts, as Popper (Popper and Eccles 1977, p. 12) puts it, that

all organisms, but especially the higher organisms, have a more or less varied repertoire of behavior at their disposal. By adopting a new form of behavior the individual organism may change its environment. . . . In this way, individual preferences and skills may lead to the selection, and perhaps even to the construction of a new ecological niche by the organism . . . thus the activity, the preferences, the skill, and idiosyncrasies of the individual animal may indirectly influence the selection pressure to which it is exposed, and with it, the outcome of natural selection.

Organic selection—or what Corning (1983) calls "teleonomic selection"—qualifies a mechanistic view of evolution, with its emphasis on the blind inexorable working of natural selection. Rather, as Corning (1983) argues, "teleonomy—that is, 'purposive' adaptation partially caused by organisms themselves—constitutes an important factor in evolutionary change," and furthermore, the influence of purposive activities "has greatly increased as species have developed greater capacity" for pursuing them. Certainly the latter observation applies with special force to the human species.

Indeed, J. H. Campbell (1987, p. 304), in terming the "relative importance of endogenous organization and external environment" to be "perhaps the central question in evolution," states that no simple answer is likely, in particular because species vary in their capacities to evolve; but as a general principle, the "genetic constitution becomes increasingly important as the species advances to a higher form." In primitive forms of life, genes are essentially passive and therefore "maximally exposed to environmental conditions"; by contrast, subsequent species "evolved increasingly elaborate ways to control mutation, suppressing deleterious types and promoting more productive types." It is the "discovery that genes are inherently dynamic instead of passive structures" that "undermine the basic Darwinian proposition that evolution is change forced on the species by the outside environment": "Species are powerfully influenced by their environment, but environmental forces are superimposed on endogenous engines for change and advancement."

Campbell (1987, p. 304) concludes:

Ever more sophisticated adaptations become potential preadaptations for developing ever more sophisticated evolutionary functions. Life asserted greater and greater control over its own evolution as it advanced; the forms progressing the fastest and farthest being those that optimized the factors important to evolutionary advancement. The increasing influence of the species in evolution not only results from advancement but also provides the mechanisms of the process of advancement. Evolutionary progress means to gain command over one's evolutionary destiny.

Thompson (1985, p. 232; cf. Schull 1990) points out that if the process of evolutionary change is "driven by internal factors," it then becomes important to address "a major gap in our knowledge of internal causes in evolutionary mechanisms": the "environmental context of these causes." In addition "to the genetic origins of particular changes," asks Thompson, "what are the ecological factors? How are the internal and external environments related?" An emphasis on the critical role of conscious choice and hence of organic selection in human evolution also suggests the increased relevance of proximate rather than ultimate causation in accounting for human action.

Sociobiological theory stresses the importance of ultimate causation, that is, explaining an activity by reference to its "adaptive significance" within the context of natural selection and evolution (see Barash 1977, pp. 37–38). Indeed, as Barash (1977, p. 38) asserts, proximate mechanisms—the immediate biological and environmental causes—can, from a sociobiological perspective, "be viewed as the servants or tools of ultimate causes, as means to an end."

Yet even Dawkins (1976, pp. 63–64) recognizes the implications of "the culmination of an evolutionary trend towards the emancipation of survival machines as executive decision-takers from their ultimate masters, the genes": "Not only are brains in charge of the day-to-day running of survival-machine affairs, they have also acquired the ability to predict the future and act accordingly. They even have the power to rebel against the dictates of the genes, for instance in refusing to have as many children as they are able to." Thus as Corning (1983) argues, an interactional process is at work in which, because proximate mechanisms may themselves influence the course of evolution, what appear to be ultimate causes may themselves turn out to be effects. In other words, natural selection is not a blind, extrinsic, and all-controlling force insofar as it represents the outcome of numerous purposive acts on the part of numerous organisms (see Schull 1990).

Proximate mechanisms, then, are not necessarily of subordinate importance in the study of human affairs. Rather, as Losco (1982) insists, both the human capacity for choice and the inevitably unique biological, historical, and cultural circumstances that attend such choice mean that an emphasis on proximate causation is especially appropriate. In fact, Losco (1982) points out that "one of the principal shortcomings of ultimate level analysis is that it does not provide a clear picture of the nature of organismic functioning upon which the social sciences can build proximate models." An evolutionary perspective is prone to bypass the biological organism, the developmental process, and the role of consciousness in accounting for our behavior—in other words, to bypass precisely the most relevant factors in understanding ourselves.

HUMAN EVOLUTION AND HUMAN HISTORY

If the study of human history is indeed, as Simpson (see chapter 4) would have it, a historical science, then it becomes subject to the same evolutionary forces that have shaped and continue to shape human evolution. In this section, I wish to relate generational and evolutionary change and, in so doing, to indicate anew their irreversible and unpredictable character and therefore their qualification of both sociobiological and behavioral science. In the next section, I will focus upon the implications of organic selection and proximate causation for an understanding of the historical process. What follows here then is in effect a Simpsonian linkage of evolutionary biology, human history, and social science as historical sciences.

Each generation differs uniquely from the preceding one, and each new generation brings with it important and irreversible changes in the art, economics, and politics of society. This truth, at once simple and universal, yet evades the canons still dominant within the social sciences (cf. Delli Carpini 1989).

From the perspective of contemporary social science, each individual is the product of external environmental forces beginning with birth. Childhood consequently becomes identified with a process of socialization whose function and end product is the creation of adults who conform to the norms and attitudes set down by the previous generation. The effective functioning of the socialization process in theory assures the continuity and stability of the major institutions in society. Generational change and conflict have little place in this perspective.

An evolutionary framework that allows for the genetic diversity of all populations and hence generations alters such a static view. Contemporary evolutionary theory implies uniquely important and irreversible generational change; and this is true independent of sociobiological theory. Trivers (1984) asserts, in the very opening lines of his classic paper on "Parent-Offspring Conflict," that "in classical evolutionary theory, parent-offspring relations are viewed from the standpoint of the parent" and hence "offspring are implicitly assumed to be passive vessels into which parents pour the appropriate care." Now what Trivers asserts certainly applies to the contemporary behavioral sciences and to their view of the socialization process; but his critical remarks are more problematic with respect to evolutionary theory. There is no question that sociobiological analysis, as Trivers has formulated it in his pioneer paper, emphasizes the presence and persistence of parent-offspring conflict, and that such an emphasis can nowhere be explicitly found in prior evolutionary thinking; but what I wish to indicate is that contemporary evolutionary thinking does, in a larger sense, imply the genetic individuality and thus independent behavior of human beings—embracing therefore both parents and children, and suggesting the necessity of generational change and conflict.

The evolutionary process, first of all, presupposes, as we have seen, both genetic and environmental diversity; for if neither the genetic composition of life nor the environmental circumstances surrounding it were ever to vary and change, no evolution of life would occur. The fact of environmental diversity and change encourages the development of genetic diversity, for life that is genetically homogeneous may find it difficult to adapt successfully to changing circumstance. It is the impact of these circumstances that provides the mechanism of natural selection in the evolutionary process.

Trivers himself (in Davis and Flaherty 1976, p. 62) observed in a conference devoted to the causes and social significance of human diversity—the year previous to the publication of his paper on parent-offspring conflict—"The genetic system not only produces the genetic variability that natural selection works on but is itself a consequence of natural selection; all the complicated mechanisms that affect variability of offspring are themselves under the influence of natural selection."

The "genetic system" in life does indeed produce genetic variability. In populations of all sexually reproducing forms of life, including therefore humans, each organism is unique. As Dobzhansky (1966, p. 57) has written of the human species: "Every human being has, then, his own nature, individual and nonrepeatable. The nature of man as a species resolves itself into a great multitude of human natures. Everybody is born with a nature that is absolutely new in the universe; and that will never appear again (identical twins and other identical multiple births, of course, excepted)."

If each human being is genetically unique, it follows that each generation of humans is unique—uniquely different from the generation before as well as the generation to follow. Each generation, after all, plays its small part in the larger evolutionary process. The advent of human culture, even the arrival of a postindustrial society with a life science that includes genetic engineering, has not halted the human evolutionary process. The title of Dobzhansky's seminal work, *Mankind Evolving*, is still apt, as well as his contention within it (1970b, p. 178) that, in the long term, the evolutionary process is irreversible—that "as evolution proceeds more and more genes are altered," so that "gene alterations" become "so numerous that the probability of retracing all the genetic steps in reverse becomes negligible" (1970b, p. 178).

Each generation does its bit to assure the irreversibility of the evolutionary process. Each generation represents an additional attempt on the part of the population to adapt to changing circumstances. The existence of generations is itself, in other words, a mechanism of adaptation. Generations exist because individuals reproduce and die. If individuals lived forever—and ceased reproduction to avoid the risk of being replaced by their offspring—there would no longer be either any generations or evolutionary process. Presumably, however, the survival of the species would be endangered: With its genetic variability thereby restricted, so also would its adaptability in the face of changing environmental conditions (cf. E. Wilson et al. 1977, p. 117); it remains of course a critically open question whether such adaptability occurs on the group or the individual level.

The uniqueness of each generation becomes significant in behavioral terms if one recalls the admonishment that Mayr leveled toward any student of human behavior well over a decade ago. Mayr (1963a, p. 650) asserted that he would be "bound to make grave mistakes if he ignores these two great truths of population zoology: (1) no two individuals are alike, and (2) both environmental and genetic endowment make a contribution to nearly every trait."

Behavioral diversity becomes a necessary end result of both genetic and environmental diversity; and one should keep in mind, with respect to the latter factor, Skinner's acknowledgment (1953, p. 424) that the environment is never the same for any two individuals. Each individual, then, with his own unique genetic endowment, his own unique central nervous system, and his own set of environmental circumstances must to some extent arrive at different opinions, even from his close neighbors or, more to the point, from his own family members. As James Madison long ago contended in *Federalist Paper No. 10*, the seeds of faction are sown in the nature of man.

This individualization of behavior, according to Simpson (1971, p. 236), has within the human species been carried "to altogether new heights" and "is a prerequisite for the human type of socialization," paradoxically finding here the "opportunity for its greatest possible development." Such a perspective hardly stresses the conformity of children to parental standards. Dobzhansky's emphasis on human genetic variability (1970b, p. 101) leads him as well to reject the behaviorism of J. B. Watson wherein a dozen babies may be made into any type of adult desired. Genetic diversity characterizes families as well as larger populations and assures both behavioral diversity and the possibility of conflict. It is of note that identical twins are significantly more cooperative and less competitive than fraternals, although there may of course be a good sociobiological explanation for this fact as well.

The developmental process in the young adds to the probability of parent-offspring conflict. Again, differences are sure to emerge on the phenotypic or behavioral level. The limited cognitive functioning of young children that Piaget describes alone ensures that parent and child will not always see eye to eye. The hormonal changes that accompany adolescence, affect behavior, and increase family difficulties are well known. Recent research increasingly emphasizes the individuality and independence of behavior in early childhood beginning from birth (see Westman 1973; A. Thomas 1976; Jackson and Jackson 1978; Kagan 1984).

The evolutionary position taken here, however, goes beyond the view that generational conflict is rooted in the nature of the developmental process of the young; that, for example, the Sturm und Drang of adolescence guarantees resistance to parental (and other) authority. In fact, this position is compatible as well with an opposing emphasis on the developmental process as one in which children are unusually malleable and susceptible to parental guidance. Some biologists indeed stress this point: Scott (1963), for example, generalizes from puppies to children in assigning to both "a critical period" during which basic social attachments are readily formed; Waddington (1960) speaks of children as being naturally "authority acceptors"; and E. O. Wilson (1975, p. 562) refers to the "absurdly" easy "indoctrinability" of humans. While it appears a matter of common experience to note the persistent occurrence of conflict between parents and their children, we must note further that generational conflict need not cease once children have matured.

Generational conflict will occur because of neurophysiological and hence behavioral diversity between generations, both within families and without. Even the most devoted young disciple will necessarily, because of a different genetic endowment, different brain, and a different set of environmental conditions, alter in some fashion the teachings of his older master. The alterations may, in fact, become so crude and vulgar as to lead the master to disavow these corruptions of his doctrine, like when Marx denied that he was a Marxist. As Dawkins (1976, p. 209) acknowledges with respect to his concept of the "meme," the cultural analogue of the gene and unit of cultural transmission, memes may not be "high-fidelity replicators at all. Every time a scientist hears an idea and passes it on to somebody else, he is likely to change it somewhat." Since all of human culture is susceptible to such alteration—which in theory is inevitable given human diversity—the lines appearing at the outset of this section should now make sense within an evolutionary framework: "Each generation differs uniquely from the preceding one; and each new generation brings with it important and irreversible changes in the art, economics and politics of society." Memes, like genes, are sure to mutate. This likelihood most assuredly extends to the memelike unit of culture, the "culturgen" postulated by Lumsden and Wilson (1981).

To the extent that it is valid, sociobiological theory further insures generational conflict. If each individual acts so as to perpetuate his own genes and each individual starts life with his own unique set of genes, conflict is inherent in social life even on the level of the family wherein parents and children will share only half their genes, or at least chromosomes (see Masters 1981). Identical twins and "supertwins" will constitute the only partial exception, that is, only in their social relationship with each other. Since they will share the same genetic inheritance they will also possess, in theory, the same underlying motivation for social behavior vis-à-vis each other. In other words, because they have identical genetic interests, their behavior ought to be unusually cooperative rather than competitive toward each other, as indeed appears to be the case (E. White 1986).

Identical multiple births aside, however, sexually reproducing life is destined, in sociobiological terms, to produce competition and conflict. As Barash (1977, p. 172; emphasis in the original) has written, "A world in which virtually every individual is genetically distinct is one in which substantial disagreements between individuals would be expected, with each selected for maximization of its *own* inclusive fitness. Therefore real possibilities exist for conflicts of interest between individuals of the same sex (males versus males and females versus females), between mates (male versus female), and even between parents and offspring." This inherent potential for social conflict—for Madisonian factional strife—is universal, then, and includes even genetically related individuals or kin and therefore extends to sibling relations (i.e., implies sibling rivalry) and parent-offspring relations. As Trivers (1974, p. 250) notes at the outset of his paper, the latter relationship, following W. D. Hamilton (1964), "is merely a special case of relations between any set of genetically related individuals."

The major theoretical contribution of Trivers's parent-offspring analysis lies in emphasizing the role of the child as an independent actor of his own vis-à-vis the parents, and thus in emphasizing the natural and continuing basis for generational conflict. Such conflict might be implicit in prior evolutionary theory, but it is not explicit. And as Trivers (1974, p. 26) rightly observes, such conflict has no basis at all in behavioral theory. A Freudian view, of course, does stress an innate and persistent generational conflict, but on theoretical grounds that remain problematical.

Trivers's analysis is inadequate, however, in treating the full range of possible conflict (and the basis for it) that may exist within families. We may see this inadequacy in Trivers's remarks concerning possible conflict between parents and offspring, even when their genetic interests overlap. Circumstances change, and as Trivers (1974, p. 262; 1981, p. 33) observes, it is the parent who is most likely to discover such changing circumstances as a result of his greater experience. Trivers does indeed extend his genetically based analysis here by pointing to the independently important role of experience. But he simply stops short of a full discussion of additional factors that may also importantly and differentially affect the perceptions of parent and offspring.

Trivers's reference to experience is ironically made within an implicitly behaviorist context; that is to say, the parent, merely by virtue of his or her greater age and life history, has the greater experience. This seems a truism, unless one reflects that mere exposure to a multitude of environmental circumstances does not necessarily insure the attainment of experience in the full sense of that word, that is, the greater capacity to adapt to changing circumstances. In short, Trivers appears to adopt here the behaviorist premise that external environmental circumstances by themselves are responsible for learning (see also Etkin 1981).

Variable environmental conditions do indeed have to be taken into account beyond genetic factors, but in conjunction with two additional factors: (1) biological development and (2) biological individuality. In the case of the former, a Piagetian genetic epistemology reminds us that children are not, biologically speaking, miniature adults and that adults are not matured homunculi. If adults persisted in viewing the world with the same brain that they had as two-year-olds, clearly little experience would be gained. This differential capacity between parent and child for gaining experience must presumably add to their potential conflict quite aside from changing environmental conditions. Parent and child are simply going to view the world differently even when they share the same interests.

Biological individuality further complicates the picture. Because of an underlying genetic diversity, all parents as well as all children will, with their unique brains, view the world differently. Thus what one parent may learn from experience another will fail to learn. Children, too, will vary in their learning capacities; in an extreme case a precocious child might even excel a slower parent in responding to a changing circumstance. Trivers, in ironic company with the behaviorists, fails adequately to treat the profound implications of human biological variability.

Trivers, in other words, in focusing on genetic relatedness, slights the dynamic and continuing interplay of diverse genetic factors with environmental ones that constitutes the developmental process and produces for each individual a set of constantly changing phenotypic traits and for an entire population a wide range of individual traits, which are in turn undergoing changes on the individual level. Such a varied and dynamic picture will likely include a broader range for conflict (both within families and without) than a strict adherence to a sociobiological position.

There is, furthermore, in the sociobiological obsession with degrees of genetic relatedness the dubious assumption of the rough equality of genes. It seems not to matter in sociobiological analysis precisely which genes are held in common. Thus there exists precisely the same basis for conflict between all parents and all children insofar as the proportion of the genes that they share is the same.

Might it not, however, make a difference precisely which genes are shared? Might there not exist a qualitative difference in genes with respect to their implications for social behavior? Certainly genes that affect such physical traits as stature, skin color, and the size of one's nose may also affect behavior; but on the whole, would not those genes that influence aptitudes and interests that may be shared in common turn out to be the most critical of all in any analysis on intrafamily conflict? Would not parents and children who turn out to share key psychological traits be less predisposed to conflict than those who shared less psychologically relevant ones, independent of the overall proportion of genes held in common?

These questions will have to await future research (cf. E. White 1986). But the next several generations of scientific inquiry into the developmental process, with particular emphasis on the central nervous system, should tell us much more when these findings are related to social behavior. A broadening of sociobiological theory in anticipation of the complexity of such findings might nevertheless be in order.

Sociobiological theory is caught between two largely incompatible goals: the desire to build a systematic, comprehensive science emphasizing the discovery of universal generalizations applying, in some cases, from termites to man, on the one hand; and an effort to explicate a biological reality, which in the very nature of things is rich, myriad, and fluid. Trivers's analysis of parent-offspring conflict provides an especially pertinent case in point.

Trivers's biology is populational, but his theoretical framework tends to be typological. Trivers is populational in his explicit recognition of important individual differences even within families. His theory of parent-offspring conflict is, of course, postulated precisely on such intrafamily genetic variation. The behaviors of parent and child will diverge because of differing genetic endowments and hence differing genetic survival interests.

In this view conflict between parent and offspring becomes inevitable. It also becomes constant and universal for Trivers, but only by virtue of his typological theoretical bent. It becomes constant and universal for any given sexually reproducing species, not in the sense that the conflict between parent and offspring takes precisely the same form in each case—for clearly it makes a difference whether one is talking, for example, about a first- or last-born in relation to the mother—but rather the same pattern obtains across the board for a particular species. This emphasis on the general, the typological, is indicated in Trivers's constant use of the phrases "the mother" and "the offspring" in his analysis, as if he were constructing archetypal roles.

An emphasis on the constancy and universality of parent-offspring conflict implies generational conflict, but not necessarily generational change. Insofar as parent-offspring conflict is constant, it recurs in generally the same form over generations. It is simply based on the recurring fact that in each generation the very same set of genetic relationships characterizes parents and their offspring. The problem that arises here is that Trivers implicitly ties behavioral diversity mechanically to an organism's degree of genetic relatedness. If, however, the unique genetic endowment of each organism, through the developmental process, also results in a unique behavioral pattern, a much more profound behavioral diversity will characterize any sexually reproducing population.

Such a behavioral diversity should, therefore, also characterize parent-offspring conflict. Perhaps partly because Jane Goodall (1971, p. 47; emphasis in the original) has, in her own words, "always been interested in the *differences* between individuals," in her description of mother-child relationships among chimpanzees she reports a wide variation in behavior—even for the relationship between mothers and their adolescent daughters where she (1971, p. 185) could only observe two cases. As against the generally easygoing and mutually protective relations between Flo and her daughter Fifi, the other relationship (whose members remain unnamed) "seemed very different: in feeding situations she was obviously afraid of her mother, and we never saw either of them show concern if the other was threatened or attacked" (Goodall 1971, p. 185).

No doubt an observer of the human family could also readily report a broad variety of behavior in parent-child relationships. To repeat an earlier conclusion, the wide range of genetic and environmental factors would seem to insure a wide range of phenotypic and hence behavioral diversity, both within and beyond the family. By stressing genetic diversity only as a function of differing degrees of genetic relatedness, Trivers's analysis takes on an overly one-dimensional quality.

Yet even Trivers's limited recognition of genetic diversity importantly qualifies the sociobiological search for universals. For it acknowledges that even within a family behavior will vary as a result of varying genetic endowments. In other words, as against Wilson's search for a species-specific biogram—a set of transcultural behaviors—and his willingness to write off cultural and group variations in behavior to environmental causes, Trivers's approach indicates that even intrafamily behavioral variation may have a genetic basis. Thus intracultural variation may also rest on some genetic basis.

Both Trivers and Wilson are, of course, entirely aware of human genetic variability. I have already alluded to Trivers's observation that the mechanism by which genetic diversity materializes has itself been selected for through the evolutionary process. What sociobiological theory, with its emphasis on universals, finds it difficult to do is to acknowledge the full behavioral ramifications of the presence of genetic diversity. For to do so would suggest an important theoretical qualification; it would mean the recognition, following Simpson, that the life (and hence social) sciences are historical in nature as against nonhistorical (as in the case of physics and chemistry).

Certainly there is continuity and universality in life, as well as change and diversity. But both of the former imply the latter. The genes now carried by all living things trace their origins to the beginning of life and are characterized by the same molecular composition, DNA. Yet the continuity of heredity is inevitably also characterized by change or mutation. And the universal operation of what Trivers refers to as the "genetic system" is such as to insure genetic variability. Insofar as phenotypic and hence behavioral characteristics are necessarily vitally affected by such change and diversity, the observer of human behavior must come to grips with this reality.

Nothing in current evolutionary science in general or in sociobiological research in particular contradicts such a populational view of life and behavior, especially for man. Yet a young, imperialistic frontier science, as it were, is impatient with subtlety, ambiguity, and qualification; and just as the behavioral sciences over a generation ran roughshod over the more complex aspects of human behavior, a sociobiological approach promises to do the same. Insofar as sociobiology constitutes an "antidiscipline" in relationship to the behavioral sciences, furthermore, its role, following Wilson (1977a), is to explain cultural phenomena by reference to their lower or biological basis and hence necessarily tends to be reductionistic in nature. A certain inevitability may inhere in this prospect. A bold new theory, however one-dimensional, must be worked through, and in the process new insights will no doubt emerge. But there is no point in denying the distortions and simplifications of the pioneering effort, for these all will have to be faced and corrected in time.

Clearly no predetermination of events characterizes either the evolutionary or the historical process. If we examine research in social science, we see that all of our studies are descriptions, however valid at the time they are made, of historical processes that, in continuing to change, ultimately refute them. This realization has begun to dawn upon social scientists. Accordingly, Nie, Verba, and Petrocik (1976, ch. 1) and Almond and Genco (1977) observe that polls and surveys do not transcend historical time and permit generalizations that can form a systematic science. As Nie, Verba, and Petrocik (1976, p. 7) acknowledge:

> The usual sample survey is a snapshot; it reveals political patterns at one moment of time. This is little problem if the snapshot is of a landscape that does not change from year to year. But if it is a snapshot of something that may look quite different next year from what it looks like today, one must use the snapshot with caution. If public attitudes—even those basic political attitudes which are supposed to originate in early life—are responsive to political events, then as patterns of political attitude and behavior discovered at one point in time they may differ substantially from that found at another point in time.

Any phenotype will reflect only the unique set of genetic and environmental circumstances that gives rise to it. As I have pointed out elsewhere (1972), a heritability estimate of a trait is only valid for the particular population studied, at the particular time that it is studied, and only for the particular trait under study. It cannot therefore be validly generalized to other populations, times, or traits. The same limitation applies, moreover, to the results of all studies in social science.

The picture of generational change that emerges from the foregoing discussion, then, is one in which such change becomes viewed as inevitable, irreversible, and unique for each generation, indeed for each family and for each individual. I wish now to indicate how even Mannheim's dynamic analysis of generational change fails fully to conform to this picture and in so doing to indicate as well how socialization studies reflect this inadequacy.

In his classic paper on "The Problem of Generations," Mannheim (1932) argued that a new generation was always in the process of emerging and making "fresh contact" with the "accumulated heritage" of the culture. As a result, the experience of each generation was novel, and generational change was inevitable. Mannheim's historical approach warns against overgeneralization based upon the experience of one generation; that is to say, if each generational experience is unique, it cannot be applied wholesale to other generations. Yet as Riley (1978) argues, many analyses do precisely this; even the sophisticated and influential work of Erik Erikson and Lawrence Kohlberg may be guilty of what Riley calls "cohort-centrism." As she cautions, "one generation's folklore and 'common sense' about the life course may no longer make sense to a later generation" (1978, p. 44).

Socialization studies may also share this problem. Thus Cutler (1977, p. 301) points out that findings which relate chronological age to differing political attitudes may be historically bound: "Why should it be assumed," he asks, "that the differences between 10- and 15-year-olds in 1960 will be descriptive of differences between youngsters who will be 10 and 15 years of age in 1980?" As Cutler might have added, we have already witnessed noticeable changes in the attitudes of American children toward political authority in the form of the president from the days of Eisenhower and Kennedy to those of Nixon during Watergate (cf. Easton and Hess 1962; and Arterton 1974). And such attitudes will, no doubt, continue to shift in the future.

Ironically, socialization studies that tend to overgeneralize their findings, and hence to take a static view by assuming generational continuity, also are typically environmentalist in their approach. For as Mannheim has indicated, on environmentalist grounds alone generations will necessarily change in unique ways. Nothing is more fluid and certain to change than environmental conditions. To attempt to build a hard, deductivist science on environmentalist foundations is to build on quicksand (Rosenberg's critique [1980] of behavioral science is in this respect well grounded).

Interestingly, the only theoretical grounds we recall (see chapter 4) for a systematic science that generalizes over time, and therefore stresses generational continuity, would be a genetic determinism that operated upon an asexually reproducing species. In other words, a human species of clones that in turn reproduced itself by cloning, and whose behavior was wholly genetically determined, would be characterized by a uniform process of socialization and the absence of generational change, indeed of politics itself—unless, of course, each clone nonetheless underwent a developmental process that involved elements of rebellious or otherwise independent behavior.

As it is, such continuity as exists between generations probably has a biological basis. Mannheim (1932, p. 319) may overstate the case when he writes that the biological principles underlying generations operate "with the uniformity of a natural law," but he does thereby point to certain biological constraints and constants. Even if the onset of puberty can and has shifted over recent generations, for example, there are nonetheless some biological parameters that govern the maturational process, ultimately limit the variation and change that is possible, and thereby make the process generally comparable over the generations. The sociobiological quest for species-specific behavior that is universal and grounded in evolution is not, after all, totally unjustified. It is the human biogram that directs the process of maturation and insures some degree of uniformity and continuity for each generation, at least in the evolutionary short run. It is this elemental human nature that also guards against an excessive historicism.

Yet Mannheim's historicist perspective does not go far enough in two key respects: It fails to take into account the reality of human genetic variability and the dynamic interactional character of the developmental process. Both of these factors importantly qualify his generational analysis. First of all, Mannheim (1932, p. 319) tends to view a generation as a uniform entity: Belonging to the same generation endows the individuals involved "with a common location in the social and historical process, and thereby limits them to a specific range of potential experience, predisposing them for a certain characteristic mode of thought and experience, and a characteristic type of historically relevant action." If we recall Skinner's acknowledgment that the environment differs for each individual, Mannheim's typological view of each generation becomes significantly qualified on environmental grounds alone. As Rintala (1968, p. 43) noted over two decades ago, "Each generation speaks out with more than one voice—there is conflict within each generation as well as among generations."

Needless to say, we must also insist on the impact of human genetic variability. As a result each generation, far from being monolithic, will include genetically unique individuals, all of whom are responding to the "same" historical events in their own unique way. To what extent members of the same generation may share historical experiences will thus always remain problematical.

The above observation applies with equal force to intragenerational factors as well. Eisenstadt's discussion (1956, 1965) of the youth groups that have emerged in modern culture, for example, must also be placed within a populational framework. Young people remain genetically and hence biologically unique during the developmental process; and this fact implies that even the most cohesive peer group will be characterized by individual variation. The proposition that such individual variation might, in part, be genetically based remains generally rejected within the contemporary behavioral sciences. Thus for the social scientist behavioral diversity is the exclusive result of environmental diversity. If biological factors are at all acknowledged, they serve only as a necessary basis for human culture and a virtually infinite variety of cultural practices. Washburn (1978, p. 412) sums up this viewpoint well: "The social sciences assume basic human biology and are concerned with the extraordinary variety of behaviors that may be learned. Human biology is primarily concerned with understanding the common biological base. The biological differences between individuals may be important in medicine and psychiatry or in achievement, but individual biological difference is rarely important to the social sciences."

Just why the study of human behavior—in contrast to the areas of human medicine, psychiatry, and achievement—should be exempt from individual genetic influences is not made clear by Washburn who, however, prefaces his remarks with the following example: Humans swim because there is an underlying biological capacity, but the particular manner in which an individual swims is a matter of learning and not of heredity; thus just as people swim in different ways that are unrelated to biological differences, people behave in different ways solely by virtue of different learning experiences. Yet Washburn concedes that human achievement may be influenced by genetic factors; is it not possible, therefore, for one individual to swim appreciably better than another in part for genetic reasons, such as genes that produce a body that is exceptionally lithe? Similarly, might not individuals be born with differing potentials and talents, with profound implications for social organization? This is not to say that each individual realizes his or her full potential, but that genetic factors may act as independently important causal agents on an individual basis. An overachiever and an underachiever may be found, for example, in the same academic environment; and while both conditions may be environmentally influenced, they may be genetically influenced as well. That a biologically literate physical anthropologist such as Washburn should discount the behavioral implications of human genetic variability well illustrates the extent of the resistance within the contemporary social sciences to the full acknowledgment of such an impact.

Now a second difficulty with Mannheim's approach is its failure to see intra- as well as intergenerational change as also real and dynamic. Mannheim tends to view each generational perspective not only in typological but also in static terms as well: Once a generational perspective becomes formed by early adulthood, it is presumed to persist through time. As Rintala (1968, p. 44) writes, "Implicit in a generation's approach to politics is the assumption that an individual's political attitudes do not undergo substantial changes during the course of his adult lifetime."

Recent studies in political socialization cast doubt on such an assumption (see Sigel and Hoskin 1977); but since none of these studies is able to separate maturational and generational factors, the explicitly biological basis for change remains unexplored. It is at least theoretically plausible that insofar as one of the defining characteristics of the phenotype is constant and irreversible change, that attitudinal and behavioral change is inevitable and rooted in the maturational process. If one grants this possibility, Mannheim's emphasis on the uniformity of biological processes through time becomes misplaced, and his analysis becomes even more dynamic than he might have wished.

The dynamic model of social change that Riley describes reasonably well accounts for the complexity of the process:

As seen in the model, the society, itself moving through time, is the composite of the several unique cohorts—at varying stages of their journey, and often traveling divergent paths. Only by understanding this fact can we comprehend the sources of tension and pressure for change inherent in the differing rhythms of individual aging and societal change. Each individual must endure continual tension throughout his lifetime because, as he learns new roles and relinquishes others, he must adjust to shifting, often unpredictable and seemingly capricious, societal demands. And difficulties can beset the society because each new cohort that is born—characterized by its own size, sex composition, distribution of genetic traits and family backgrounds—requires continual allocation and socialization for the sequence of roles it must encounter within the prevailing social structure. Small wonder, then, that there can be no fixed process of aging! (Riley 1978, p. 42)

Nonetheless, there is one fairly important question raised by the Riley model that invites special examination by those interested in the political socialization process. This question concerns the part played by that process in social change. Riley correctly points to the difficulties that beset society because each new cohort "requires continual allocation and socialization for the sequence of roles it must encounter" and because, therefore, each individual "must adjust to shifting, often unpredictable and seemingly capricious, societal demands." What is missing in this formulation, however, is precisely the emphasis provided by Trivers's parent-offspring analysis, wherein all parents and all children have their own individual genetic interests, and therefore in theory at least can be expected to assert them actively—even if these run counter to parental or other social authority in the case of the children. The aims of the socialization process and the actual outcomes in short may turn out to be very different; and the process of socialization becomes, then, as dynamic as the model Riley proposes for social change: Indeed, if one reflects upon it, the two processes are intimately related. As I have insisted elsewhere (1969, 1979, 1980a, 1981c), the attempt by adults to socialize children must be conceptually distinguished from the act of learning on the part of children. Learning is a dynamic and purposive process. According to Granit (1977, pp. 74–75), "learning complex matters presupposes conscious awareness," which in turn "makes possible long-range anticipation of events" that "is tied to purposive acts." This evolutionary adaptive and innate capacity means, as Popper (in Popper and Eccles 1977, p. 158) puts it, that we "are highly active problem-solvers."

If learning is a dynamic process on the part of children, the use of the term *socialization* should be restricted to the attempt of adults to obtain conformity from children to their norms; that is to say, what adults attempt to socialize and what children in fact learn often differ (cf. E. White 1969). To say that children are socialized is to imply their passive conformity. To say that they learn is to acknowledge their active participation. Therefore, those who wish to emphasize the attempt of adults in society to impose their views on the younger generation should refer to their area of concern as socialization, whereas those who wish to focus on how the child acquires his views on government and politics will identify their field as "political learning."

Because learning in particular and conscious awareness in general are goal directed and represent the effort of the individual to select and even to create optimal environments, studies in political learning must acknowledge what Darlington (1969) refers to as self-selection. As Langton (1969, pp. 116–17) has pointed out, "an important part of the difference in political orientation between those from different levels of education, which is frequently cited in the literature and is usually explicitly or implicitly ascribed to the 'education process' may actually represent a serious confounding of the effect of selection with that of political socialization."

Of course, oppressive environments do oppress. But even they may not always succeed in their socialization attempt. Frederick Douglass was sent to a Negro breaker in order that his spirit be broken. Yet he not only survived the experience in body, but he also triumphed in spirit. Following a knockdown fight—a draw but a moral victory—with his master, Douglass (1963, p. 74) wrote that he regained his "departed self-confidence." Indeed, it had been that self-confidence that caused him to be sent to the Negro breaker in the first place and that had permitted him, despite a hostile environment, to teach himself to read and write. While the slave environment could largely (although not wholly) control his behavior, it could not control his inner thoughts. For Frederick Douglass—and for children generally—the attempt to socialize and the act of learning are not the same. Ultimately, the process of learning, and the continuity and the change that it implies, will be at least in part explicated by an emerging human neurobiology.

If behavioral theory is clearly too often guilty of viewing the socialization process in passive, static, and unidirectional terms, sociobiological theory is itself somewhat schizoid in this matter. Wilson (1975, p. 56; emphasis in the original) writes, after all, that "human beings are *absurdly* easy to indoctrinate—they see it." While a high degree of genetic relatedness characterizes parent-offspring relations and therefore suggests cooperation as well as conflict, acceptance as well as resistance, nonetheless Wilson's statement would appear to go well beyond the mixed and ambiguous nature of a relationship implied by a 50 percent sharing of genes. Of course it may well be that young children in particular are unusually receptive to adult direction, that they both seek and need it, although even here individual difference of some import may well be present. It is noteworthy, in any case, that the two major concepts that Wilson borrows from social science—"role" and "socialization"—carry with them into sociobiological theory the same denial of an active self-awareness that is found in their source.

I would only mention the possibility that at least some apparent instances of indoctrination are successful on the surface only, that outward conformity does not always imply inner acceptance. In other words, children (or for that matter, adults) may be deceitful in their relations with authority.

To provide an instance: A graduate student in political science who as a youth had escaped Hungary during the 1956 revolution once told me of his participation in the rites marking the death of Stalin in 1953. It seems that he and his adolescent peers marched in this ostensibly solemn ceremony, but privately wisecracked among themselves; and while they apparently had all they could do to keep from "cracking up," they nonetheless managed to appear properly respectful to an outsider. Thus from a distance, order and decorum were preserved; closer in, an irreverence was present that was camouflaged for obvious reasons and that proved to be quite real, as the revolutionary events of 1956 were to show.

The peaceful revolution of 1989 that swept Eastern Europe provided even more graphic demonstration that private beliefs, allowed free expression, can swiftly take public form, replacing wholesale a previously forcibly imposed social system. Totalitarianism is "something that is invisible at a distance," Vaclav Havel wrote in 1987 (cited in Brzezinski 1988, p. 111): "In our system, the violence is spiritual rather than physical. In other words, hidden, covert. Life here seems pretty normal to outsiders." But, Havel continues, those who are "at the mercy of the all-powerful bureaucracy," live "their lives in a state of permanent humiliation."

Clearly, though, the humiliation proved to be less than permanent. Writing a postmortem on the decade of the 1980s in the *New York Times Week in Review* (December 24, 1989), R. W. Apple, Jr., concluded that "if the 1980s proved anything, they proved once again the power of ideas." Not least was this true for Eastern Europe in 1989, where the concentration of formal power in the Communist party and the military simply collapsed when, because of Gorbachev, it was no longer propped up with Soviet tanks.

"True believers" surely exist; but not every member of a youth group (or other organization) matches outward conformity with inward belief. As the Russian futurist poet Mayakovsky (cited in Solzhenitsyn 1974, p. 42) wrote:

> Think
>   about the komsomol
>   for days and for weeks!
> Look over
>   your ranks,
>   watch them with care.
> Are all of them
>   really
>   komsomols?
> Or are they
>   only
>   pretending to be?

The possibility that an individual's inner beliefs and outer behavior may diverge speaks to the need in social science research to supplement mass survey research projects with intimate, in-depth studies. The latter may not effectively resolve this question, but the former clearly will not. One should note that Trivers's emphasis (1981) on the possibility of deceit as a strategy of behavior, especially as it might be placed within a conscious or intentional framework that would begin to apply to adolescence, suggests that not all the findings of socialization (or any other) research can be taken at face value.

It remains to reiterate the necessity of placing social science studies fully within the context of genetic and environmental diversity, that is, recognizing the biological and historical parameters that limit each study, and thus warn against generalizing beyond them. Both the behavioral and a narrow sociobiological approach share this propensity for overgeneralization. A narrow sociobiological approach, as I have written elsewhere (1978) tends to focus unduly on inclusive fitness theory as the basis for all behavior, whereas a broad approach leaves rather tentative and open the basis for behavior as it attempts a synthesis of the continuing scientific research undertaken in such socially relevant fields as ethology, population genetics, and neurobiology.

The "generation gap" provides a case in point. Trivers's analysis implies constant conflict between the generations, but does not concern itself with cultural and historical factors that might significantly affect the nature of that conflict. Mead (1978), on the other hand, sees a qualitative progression from "post-figurative" through "co-figurative" to "pre-figurative" cultures. In the former, "children learn primarily from their forebears"; in the intermediate, "both children and adults learn from their peers"; and in the latter, "adults learn also from their children" (Mead 1978, p. 2).

"Post-figurative cultures," writes Mead, "in which the elders cannot conceive of change and so can only convey to their descendants this sense of unchanging continuity, have been, on the basis of present evidence, characteristic of human societies for millennia or up to the beginning of civilization." Mead notes, however, that "intergenerational relationships within a post-figurative society are not necessarily smooth." Certainly parent-offspring conflict of the type Trivers describes is not incompatible with this model.

Prefigurative culture, on the one hand, is the result of irreversible changes that have taken place since the beginning of the industrial revolution. "The primary evidence that our present situation is unique, without any parallel in the past," Mead asserts, "is that the generation gap is world wide" (Mead 1978, p. 18). Yet Adelson (1979, p. 33) concludes, writing of the 1960s that formed the backdrop for Mead's work, "There was, in fact, no generation gap, at least no more so than in any other historical period, so far as one can tell from the evidence."

These divergent views have been raised, not in any attempt to resolve them but rather to indicate the problems that arise when a historical finding is overgeneralized. Mead, for example, may have overstated the extent of youth rebellion based on the student protest movement of the 1960s. Adelson rightly points to the likelihood that atypical adolescents, such as those involved in student protest movements, monopolize both scholarly and popular attention, with the result that their behavior becomes overgeneralized to encompass an entire generation. Yet Adelson, too, may be overgeneralizing in stressing the similarity of intergenerational relationships over historical time. Mead's argument for qualitatively irreversible changes is compelling; and as she herself has noted, such irreversible cultural evolution is the analogue of evolutionary change.

If cultural change is, in fact, irreversible, then both a narrow sociobiology and a deductivist social science need to be corrected in order to place their findings within a unique historical and evolutionary context. Empirical findings, however valid, may constitute no more than a snapshot of what is otherwise a part of a unique and dynamic process; and longitudinal studies, unless perhaps continued indefinitely, provide no definitive solution to this problem. Continuous longitudinal studies, moreover, are at this point premature, since for obvious reasons these should not be undertaken before the social sciences have accepted a theoretical approach that can, with some reasonable assurance, be projected into the more distant future. Neo-Darwinian evolution has served this purpose for biology for over a century and promises to do so indefinitely, even if with qualification. Now such a perspective is quite likely to be extended to the social sciences as well, but not so much in the manner that a narrow sociobiology might envision, but rather, as Simpson has argued, to join both the life and social sciences together as historical studies. This development would insure that both social scientists and sociobiologists cease chopping up a complex and dynamic reality to fit a procrustean theoretical emphasis on uniformity and predictability.

ORGANIC SELECTION AND THE HISTORICAL PROCESS

We may recall in our discussion of organic selection that the organism may play an active role in the evolutionary process. This possibility is especially likely in the case of humans uniquely endowed with conscious foresight. Such human characteristics are just as likely to affect the historical process. To establish more clearly the importance of the human organism in general within the historical process, let us imagine three hypothetical societies: the first comprised only of idiots; the second only of "Einsteins"; and the third comprised of all varieties of individuals— excepting genius.

Let us return to the "noncultural retardates" that Wilson, following Wills (1973), refers to, and imagine a society exclusively populated by such individuals. One may recall that the capacity for a distinctly human language is here denied. The result would be as if human evolution had been swept back a million years to the protoman of presymbolic speech—or as if human ontogeny ceased at age two. The major practical consequences are clear. Perhaps the most obvious is that human culture as we know it would end. For the arts and sciences, religion and education, commerce and politics, all rest on the capacity for abstract thought and speech. And that capacity in turn rests firmly upon a biological foundation. Man through his culture does not transcend his nature, but rather absolutely requires it for his uniquely human achievements.

At the other extreme, we may ask the question raised by Etkin and echoed by Mayr (1963a) as to why human evolution did not ultimately produce a population of Einsteins. Why does the potential for human diversity not become actualized as genius? In its most extreme form, the question becomes "why all animals are not as intelligent as Einstein and as moral as Schweitzer" (Etkin 1954, p. 130).

Etkin argues that selection pressures favored those protohominids who could participate in monogamous family units and future-oriented hunting bands, both of which demand the capacity for stable cooperative relationships and symbolic communication, capacities largely out of the biological reach of all other animals, on the one hand, and yet capacities not at all requiring the highest abstract thought, such as "motivational reasoning," or the highest ethical behavior, such as is implicit in "the golden rule." These higher capacities, according to Etkin (1954, p. 140), "are not explicable in the biological terms developed here."

Yet whatever the biological explication of human genius (for a genetically based explication, see Lykken [1982]), it can and does occur. Indeed, its uppermost limits have perhaps hardly been broached. "If, for example, Mozart's constitution had contained the maximum concentration of genes favoring musical genius of anyone yet known," Crow has written, "this does not by any means imply that his was the best genotype that could have been constructed from the pool of genes available in a large population" (1961, p. 427). Let us imagine, accordingly, that all reproduction in a given generation tends to realize "the best genotype" possible along various lines of excellence. What would be the practical outcome?

No doubt a society would emerge with a capability for far-reaching achievements that exceed anything man can at present envision. This author, at least, is in no position to discuss the possible contribution to theoretical physics by a super-Einstein. It is, however, possible to detect certain negative consequences resulting from a human community populated solely by geniuses.

A society of geniuses, whether evolved naturally through an improbable biological reproduction or artificially through cloning, would destroy the basis for an essential division of labor. Human stratification and mobility reflect in part, however imperfectly, an underlying genetic variability, and also serve to perform the vast number of tasks that must be accomplished if contemporary urban societies in particular are to survive. Addressing himself to the possibility of producing a uniform human being through cloning, Ravin (1972, p. 15; emphasis in the original) concludes: "Genetic heterogeneity may assure survival and adaptation *of the species as a whole* in a discontinuous environment. We may need, in short, to cherish the genetic variations between individuals." (I leave open here the question of individual versus group selection.)

What if human societies evolved with all manner of men—with the sole exception of genius, which, as Etkin observed, may not be biologically explicable in the context of the demands of protohominid life?—Which may be precisely the point: Human life without genius is the life of prehistoric man. According to Mayr, "J.B.S. Haldane and others have pointed out that the human achievements—most things that distinguish men from the animals and are responsible for man's civilization, art, literature, and science—were achieved by less than 1 percent of the human population, by those in the upper tail of the curve of human variation in inventiveness, imagination, perseverance, and ability to think clearly" (1967, p. 836).

The argument stressing the role of the environment in multiple inventions and discoveries (cf. Kroeber 1948; L. White 1949) usually overlooks the obvious fact that, for example, both Leibniz and Newton, the coinventors of the calculus, were geniuses and that if neither had existed the calculus might have had to wait another generation. On the other hand, the time was not ripe for the discovery of the principles of genetics when Mendel proceeded to discover them anyway—and his discovery had to wait another generation before making its full impact (cf., however, Mayr [1982, p. 722]). As Eccles (1973, p. 222) insists, "Geniuses are born, not made," even if "they have to find the particular metier that matches their exquisite brain potentiality": "We still are gravely underestimating the tremendous range in the brain performance apart altogether from environmental influences."

Mozart's genotype may not have been the very "best" for "favoring musical genius," but it was nonetheless uniquely extraordinary in this respect. He might have died even younger, during early childhood as once appeared likely, but now we are aware of his unique and irreplaceable artistic contribution. Similarly, if Beethoven had committed suicide at age thirty as he had contemplated, the world could never have duplicated the music that he was to compose in the succeeding quarter-century. As it is, Mozart's untimely death at thirty-five has meant a loss whose magnitude we can never fully gauge. In either case, the losses in question would have been correspondingly greater if individuals with their particular genotypes had not materialized in the first place. "If one can imagine a world of music without the literature of Mozart," Kolodin (1980, p. 312) writes, "one can imagine English literature without Shakespeare." In general, then, there can be no music without musical talent and no exceptional music without exceptional talent. A similar observation applies to every human endeavor.

In political affairs, for example, a vigorous Lenin living well past his allotted fifty-four years could not fail to have changed the course of Russian history in ways now unforeseen, but the mere fact that Lenin lived at all changed the course of world history. Such at least is the contention of Hook (1955), who sees Lenin as an "event-making man." According to Hook, the Bolshevik Revolution would not have occurred without the leadership of Lenin. And of course Lenin could not have developed without his unique constellation of genes—albeit acting always in conjunction with environmental factors.

Similarly, if Hitler had not been born, Nazi Germany may not have emerged. For what other Nazi leader could have duplicated Hitler's rise to power? One may observe that when George Wallace was incapacitated following an assassination attempt during the 1972 presidential campaign, no other political leader, including notably Lester Maddox, could capitalize on his demonstrated popular support. The only factor that had changed involved the personal effectiveness of George Wallace. As Carlyle (1956) has argued, the "times" as such do not throw up leaders, for if the potential for leadership (however perverted) is not there, the "times" are powerless to evoke it. It is, moreover, precisely the case with the most exceptional leaders that individual differences become amplified and therefore especially significant independent influences of their own. Bell (1982, p. 51) has observed that "social science questions often hinge on particular situations" that involve "the will and character of individuals." He goes on to speculate that without Charles de Gaulle at the helm in 1958 in France, "the revolt of the French Army in Algeria against the French government might have taken a successful course." We might extend the thought experiment of Popper (in Popper and Eccles 1977; see ch. 3) involving brain transplants to world leaders. What, for example, might have been the consequences if the brains of Winston Churchill and Neville Chamberlain (or Adolf Hitler!) had been somehow exchanged in 1938? And lest genetic factors be viewed as unimportant here, imagine the 1938 brain of Churchill replaced by one of a severely retarded individual, a victim of a genetic defect. Any number of similar conjectures at least suggest the importance— and the uniqueness—of individual will and intellect in great leaders. Individuals can and often do affect the course of history.

Hook (1955), in restricting such influence to only a handful of "event-making" leaders, may in fact underestimate this impact; for he relegates all other leaders to a largely passive category, which he terms "eventful." Tucker (1981, p. 30) rightly protests this rigid division:

For good or ill, leadership influences events. Leaders are hardly divisible into the few "event-making" ones who impose their personalities upon history and make it go their way and, on the other hand, the "eventful" ones who only help it follow the course it was going to take anyhow. The ways in which leaders count cannot accurately be expressed in such a dichotomy. They matter in degree, a little more, a little less, depending on how they diagnose those problem situations for their political communities, what responses they prescribe for meeting them, and how well they mobilize the political community's support for their decisions. The difference between the way in which one leader performs these functions and the way in which another does may not be very great. But it is not unimportant: it matters in history.

If individuals can influence events as well as be influenced by them, then it makes sense to study especially those individuals who are most influential. This is precisely what I propose in the following chapter.

6

Self Direction and Political Action

Conscious Purpose, Emic Analysis, and Ongoing Political Biography

Sympathy is your real key to the riddle of life. If you can put yourself in men's places, if you can see the same facts from the points of view of many scores of men of as many different temperaments, fortunes, environments, if you have Shakespearian range and vision, then things fall into their places as you look upon them and are no longer confused, disordered, scattered abroad without plan or relation. You must not classify men too symmetrically; you must not gaze dispassionately upon them with scientific eye. You must yield to their passion and feel the pulse of their life when you are studying them no less than when you are acting for them.

—Woodrow Wilson (1911)

Social scientists who—for whatever philosophical or methodological reasons . . . view human behavior as simply reactive and consequently susceptible to the same explanatory logic as "clocklike" natural phenomena are trying to fashion a science based on empirically falsified presuppositions. That becomes clear when their explanatory schemes are thought of in terms of their own behavior as scientists. Insofar as they acknowledge the importance of scientific memory, scientific creativity, calculative strategies, goal seeking, and problem solving in their own work, they must in some degree acknowledge these qualities in the human and social material they investigate and seek to explain.

—G. Almond and S. Genco (1977)

Just as I write this chapter with the intention of influencing the reader's thinking along certain lines, and I assume that the reader is capable, if not willing, to be so influenced, I assume that similar sets of relationships involving intentions and capabilities pervade the realm of politics. More specifically, I assert the following: First, that human conscious awareness and purposive action are biologically adaptive as well as culturally influenced capabilities. Such evolved capabilities, moreover, may transcend both their biological and their cultural determinants and, as emergent forces, act as causal agents that motivate behavior. Individuals may not always achieve their ends, but to the extent that they do, the outcomes may be said to reflect a process of individual self-selection; and even failed objectives may be viewed from the starting point of what the individual was attempting to do.

Second, if conscious purposes may in fact act as causal forces, then the force of human self direction—as against genetic or environmental direction—becomes theoretically acknowledged. It follows, third, that emic modes of analysis are most appropriate for the study of human action. Emic analysis focuses upon the meanings that the actors themselves impute to their actions; and such a perspective becomes especially needed in the case of unusually creative individuals. Weberian *verstehen* is the hallmark of such an approach.

I apply the foregoing analysis, finally, to political action and leadership. We here enter the fledgling field of neuropolitics (see Davies 1976; Schubert 1981a, 1981b, 1983, 1989; Hines 1982; Losco 1982; Masters 1989; Peterson 1982; and E. White 1982a); and within the context of emic analysis and political action, I resurrect a proposal made a decade ago for political scientists to engage in ongoing political biography (White 1972) in which political observers follow major contemporary political leaders throughout their careers in a comprehensive manner that includes a critical consideration of their inner motives and expressed objectives.

SELF-SELECTION, EVOLUTION, AND CONSCIOUS PURPOSE

Each organism, according to Darlington (1969, p. 269; cf. Scarr and McCartney 1983), seeks and, if possible, selects "the environment that fits it best." Such self-selecting activity is both adaptive and purposive. It is also pervasive within the animal kingdom. "The occurrence of goal-directed processes," writes Mayr (1976, p. 389), is perhaps the "most characteristic feature of the world of living organisms." It includes, Mayr states, "most activity connected with migration, food-getting, courtship, ontogeny, and all phases of reproduction." Mayr terms such goal-directed activity "teleonomic," by which he means that it is both fully and internally caused. In ascribing teleonomy to organic life, Mayr (1976, p. 402) endorses what he refers to as the Kantian insight that "no explanation of nature is complete that cannot account for the seeming purposiveness of much of the development and behavior of living organisms."

The instinctive behavior of lower forms of life is probably just that,

Griffin's speculative ascription of conscious awareness to honeybees (1976) notwithstanding. But the existence of such awareness in higher, though intrahuman, forms is at least problematic; and, as Griffin (1976, p. 104; 1984) argues, it would confer "a significant adaptive advantage by enabling animals to react appropriately to physical, biological, and social events and signals from the surrounding world." For Jerison (1973, p. 17), "biological intelligence may be nothing more (or less) than the capacity to construct a perceptual world. For man, this is the real world of which he is conscious."

Conscious awareness in humans needs only to be experienced to be demonstrated. Its adaptive value seems clear. "Conscious awareness has been an enormous asset in the evolution of higher organisms," Granit (1977, pp. 75, 76) writes. "Its raison d'être is perfection of control over the environment." Consciousness, Granit (1977, p. 74) further points out, "makes possible long-range anticipation of events." As Dobzhansky (1967a, p. 56) illustrates, "to make a tool for a future employment one needs more than mental dexterity; what is necessary is formation of a mental picture of a situation which is expected to arise in the future, but which is not yet given to the senses."

P. S. Wilson (1980, pp. 100–101), arguing that "a conscious sense of purpose is a variant selected for because of its adaptive advantages for us as a species," concludes that it "is itself a species characteristic." If this view is valid, an observer would do well to consider the critical role that conscious purpose might play in human affairs (cf. Itzkoff 1983, 1985, 1987).

CONSCIOUS PURPOSE, NEUROBIOLOGY, AND HUMAN ACTION

The fact of conscious purpose is an indisputable datum of our own everyday life. Yet with the rise of behaviorist theory, as Pribram (1981, p. 154) observes, "the intentional and intensional aspects of cognitive processing were not only ignored but considered inappropriate for scientific analysis." In recent years leading brain scientists, Pribram included (see chapter 2; Pribram 1976b; Sperry 1980, 1986; Popper and Eccles 1977; Eccles 1989; and Granit 1977), have begun to treat the neurophysiology of conscious awareness and purpose and to support its autonomy and causal potency as an emergent property of the higher level organization and functioning of the central nervous system. Although important differences exist in detail and concept in the approaches of these neuroscientists, their central conclusion remains, we recall, essentially the same: Conscious purposes may themselves constitute causal agents that are not reducible as mere epiphenomena of genetic, physiological, physiochemical, or environmental factors that otherwise determine brain functioning.

An emphasis on the causal role of conscious purpose is especially important for the scientific study of man. In psychology there is now under way, in opposition to the once dominant behaviorist theory, what the Kreitlers (1976, p. 3) refer to as "the revolutionary comeback of cognition" (see also Pelletier 1978; Wicklynd 1979; Hilgard 1980; Hunt 1989). This conceptual shift has also affected social psychology: "The major, almost revolutionary recent change in psychological social psychology," writes Stryker (1981, p. 387), "is that the subjective has become respectable." The subtitles of three recent books indicate this shift: Jack Katz, *The Seductions of Crime: The Moral and Sensual Attractions of Doing Evil* (1988); Dorothy Tennov's *Love and Limerance: The Experience of Being in Love* (1989); and Mihaly Csikszentmihalyi's *Flow: The Psychology of Optimal Experience* (1990). Each explicitly highlights the centrality of the subjective experience, whether in crime, love, or happiness.

Now if our subjective thoughts necessarily occupy the present, they frequently are directed toward the future. "At any moment in any man's working and conscious life," writes Hampshire (1959, p. 169), "there is always a set of possible true answers to the question—'What is he doing now?' For human beings, to be conscious is to have active intentions." This reality in turn implies, as Etkin (1981, p. 67) points out, that "a completely mechanistic interpretation of human behavior is impossible because feedback from the anticipated future modifies the present nexus of causes and changes them."

If, as Midgley (1978, p. 65) insists, motives must be taken seriously, the focus of the social sciences shifts dramatically. As we recall from chapter 3, the behavioral sciences, with their emphasis upon overt behavior as the unit of study, become transformed, following Parsons (1978, p. 5) who specifically includes psychology within this classification, into the sciences of human action. The concept of action in its Weberian formulation (see Weber 1947; Reynolds 1976; E. White 1981a) embraces both the outer behavior of an actor and its inner meaning, that is, the motives that appear to impel it. If we acknowledge the validity and hence the necessity of refocusing our attention from behavior to action, we must consider the appropriate methods and analysis consistent with this refocus.

ETHOGENY, EMIC ANALYSIS, AND SELF-DIRECTION

The chief task of the observer of human action is to discover the inner meanings and motivations that give rise to overt behavior: "At the heart of the explanation of social behavior is the identification of the meanings that underlie it." Harré and Secord (1972, p. 9; emphasis in the original), who refer to this discovery and identification of such meanings as "ethogeny," point out that an important part of this approach "involves the obtaining of *accounts*—the actor's own statements about why he performed the acts in question, what social meanings he gave to the actions of himself and others."

Ethogeny presupposes emic analysis. Both emic and etic approaches grew out of studies in linguistics, but have been extended to human culture as well as language (Pike 1967; Harris 1968; Crook 1980). The manner in which Pike (1967, ch. 2), the linguist who coined the terms, defines, differentiates, and applies them to both culture and language may be summarized as follows: The emic standpoint views behavior as it originates from within a culture, while the etic studies it from the outside. Thus the emic observer (possibly a participant observer) attempts to learn the meanings that the members of a group or culture themselves impute to their behavior. An etic observer, by contrast, in effect imposes his own systematic perspective (whatever it might be) onto any particular group or culture.

A number of further distinctions, among others, result: An etic approach tends to be broad, systematic, and comparative, treating all or many cultures and languages at the same time. An emic view, on the other hand, focuses on one specific culture or language. An etic approach thus selects, segments, and abstracts, while an emic one stresses a total wholistic picture.

The etic constitutes, for Pike (1967, pp. 37–39), "an essential initial approach to an alien system," but "the initial etic description gradually is refined, and is ultimately in principle, but probably never in practice, replaced by one which is totally emic." In an etic approach, the plan, the criteria, and preview of the study are all external to the units under observation; whereas in an emic approach the reverse is the case: Those who are most familiar with a way of life or speech—those situated within the community—define the relevance of the concepts and data.

The etic plan of study is created; the emic involves discovery: "The etic organization of a world-wide cross-cultural scheme may be created by the analyst. The emic structure of a particular system must, I hold, be discovered" (Pike 1967, p. 38). Pike might have added that from the emic standpoint, the internal structure of a culture or community is itself a creation of its members.

Although in principle the etic and emic approaches represent polar and exclusive plans of study, in practice they may be useful and complementary modes. A comparison of cultures must necessarily be transcultural, but an understanding of a particular culture must include the reported understandings of its members. I in effect asserted the complementarity of the two approaches elsewhere (1981a) when I argued the desirability of counterpoising and coordinating mass surveys and small in-depth studies; for instance, Stouffer's *Communism, Conformity and Civil Liberties* (1955) and Smith, Bruner, and White's *Opinions and Personality* (1953) might usefully have been conducted in conjunction with one another.

Indeed, the most glaring fallacy is the insistence that one approach—the etic or emic alone—is sufficient for analysis. An emphasis on an etic to the exclusion of an emic approach is, however, all too often the result of a desire to attain a "hard," systematic, objective science. Thus what is holistically unique and by definition esoteric necessarily escapes detection. The dangers of an exclusively etic position were anticipated by Sapir years ago (cited by Pike 1967, p. 38; Pike's interpolations are found within the brackets):

It is impossible to say what an individual is doing unless we have tacitly accepted the essentially arbitrary modes of interpretation that social tradition is constantly suggesting to us from the very moment of our birth. Let anyone who doubts this try the experiment of making a painstaking report [i.e., an etic one] of the actions of a group of natives engaged in some activity, say religious, to which he has not the cultural key [i.e., a knowledge of the emic system]. If he is a skillful writer, he may succeed in giving a picturesque account of what he sees and hears, or thinks he sees and hears, but the chances of his being able to give a relation of what happens, in terms that would be intelligible and acceptable to the natives themselves, are practically nil. He will be guilty of all manner of distortion, his emphasis will be constantly askew. He will find interesting what the natives take for granted as a casual kind of behavior worthy of no particular comment, and he will utterly fail to observe the crucial turning points in the course of action that give formal significance to the whole in the minds of those who do possess the key to its understanding.

Thus far our discussion of emic and etic approaches has assumed a culture or group to be the unit under study. What, however, of the individual? Should individuals be subject to either or both emic and etic analysis? At this point, we would do well to return to the points raised in the previous section on conscious purpose, for the answer will depend critically on key assumptions made concerning the wellsprings of our behavior.

Three opposed sets of assumptions can be made: First, that we and our behavior are under genetic direction; second, that they are under environmental direction; and third, that they are self-directed. The latter two possibilities are described by Harré and Secord (1972, p. 8) when they describe two formulations as "(1) the person acting as an agent directing his own behavior, and (2) the person as an object responding to the push and pull of forces exerted by the environment." The former, they note, emphasizes "self-direction," the latter environmental contingencies. Now if the authors were to have revised their work a decade later, they might well, it is plausible to speculate, have included the third possibility of some form of genetic control. In any event, let us examine (1) genetic direction, (2) environmental direction, and (3) self-direction in relation to the relevance of etic and emic plans of study. (Clearly an interactional approach is not only possible, but also desirable [cf. E. White 1972]; but for my present purposes these alternatives are presented starkly.)

Genetic Direction

The recent emergence of sociobiological theory makes it the clearest exponent of this formulation. Its central theorem has been summarized by Barash (1977, p. 63) as follows: Organisms "should behave so as to maximize their inclusive fitness," the latter being defined as the representation of the organism's own genes and of copies of those genes in kin within future generations. In other words, whatever individuals may think they are doing, their behavior is in fact best explained by underlying genetically based considerations. Indeed, Trivers (1981) has emphasized how much of our behavior may be based upon unwitting self-deception as a result of such underlying considerations. In any case, sociobiological theory, with its emphasis upon revolutionary or ultimate as against proximate or organismic causation, imposes its explanations upon its subject matter. As Crook (1980, p. 173) observes, "the attempt to explain certain cultural phenomena through recourse to sociobiological theory is etic."

The etic nature of sociobiological explanation is clearly shown by an examination of Lumsden and Wilson's *Genes, Mind, and Culture* (1981). The central unit of analysis in this work is the "culturgen," which refers to any of "an array of transmissible behaviors, mentifacts, and artifacts" within culture (Lumsden and Wilson 1981, p. 7). The culturgen, upon examination, turns out to be a general, uniform, and static concept much like Dawkins's meme (Dawkins 1976, ch. 11), which Lumsden and Wilson explicitly state is an equivalent unit. Thus what I have elsewhere (1978, p. 281) said in criticism of the meme applies equally to the culturgen: that it does not adequately take into account the dynamic development process, which includes an ever-changing Jamesian stream of consciousness wherein no thought of our own remains wholly intact; nor does it fully realize that it is an illusion, as Sapir (1956, p. 204) pointed out long ago, to view culture as a "neatly packed up assemblage of forms of behavior handed over piecemeal, but without serious breakage" to the newest generation.

The concern of Lumsden and Wilson, as I asserted in chapter 1, to build a quantitative, systematic, and predictive science means that unique individuals and irreversible change—the very stuff of organic evolutionary life—become overshadowed. The "mind" in the title of the work is found only in populations as a statistical entity; it has no personal embodiment.

It is of note that in recent years several sociobiologists have begun to criticize an unduly etic approach. Crook (1980, p. 173; emphasis in the original) writes, for example, of the dangers that arise "when etic theories of a highly reductionist kind eliminate the consideration of emic material": "Certain models of human interaction concentrate entirely on the process itself without examining the quality of the personal aspects. In such studies the *meaning* of such meanings to the people concerned often remains unanalyzed. Sociobiologists can avoid the charge of reductionism by retaining an awareness of the multiplicity of levels responsible for human action" (cf. Symons 1979, pp. 206–7).

Wrangham (1980, p. 176) also critically evaluates sociobiology as follows:

To the extent that novel behavior patterns are found to be employed adaptively, sociobiology will be forced to change. Individual choice will become a reasonable alternative to natural selection as the immediate determinant of species' behaviour patterns. Genetic questions will then be referred away from particular acts towards more complex characteristics, such as those concerned with maturation, emotion, learning abilities and insight. As this happens, the grip of genetic determinism will be relaxed, and sociobiologists will feel freer to pay attention to individuals and their private minds.

Environmental Direction

When Harré and Secord published *The Explanation of Social Behavior* in 1972, they noted (p. 9) that the "most typical predilection" in the behavioral sciences was for "environmental contingency explanations." The most extreme form of such explanations is to be found in behaviorist psychology; and also in 1971, Skinner was to publish his *Beyond Freedom and Dignity* in which he asserted (p. 205) that "an experimental analysis shifts the determination of behavior from autonomous man to the environment—an environment responsible both for the evolution of the species and for the repertoire acquired by each member."

Of course, environmental direction can in theory take different forms; cultural and economic determinism are especially familiar. Leslie White (1949, p. 184) asserts, for example, "that the human mind is a function of the cultural system that embraces it. What it does, what it believes, thinks and feels, are determined not by the individual, but by the circumambient culture." What all theories of environmental direction imply is the devaluation of individual autonomy and therefore the inappropriateness of an emic approach. Since they all necessarily postulate that the individual is directed by some external set of contingencies beyond his control they necessarily admit an etic perspective: The behavior of the individual or the group is to be explained by reference to factors extrinsic to them.

It is now clear that theories of both genetic and environmental direction necessarily employ etic plans of study, however different their explanatory content may be. The attraction of such an approach is clear: it permits an objective, systematic, and predictive science to develop. At least it does so in principle. For it should be noted that even if a particular etic approach were wholly valid, it might not be able to deliver the scientific results that it promises. The primary reason is that an understanding of the factors that direct behavior does not insure, unless one can also control the factors, that one can predict that behavior—any more than an understanding of the conditions underlying the weather has led to successful long-range forecasts. Thurow (1977, p. 87) makes this point graphically in relation to economics:

There is ... a fundamental confusion between economic understanding and the ability to make economic predictions. Most physical sciences understand rather than predict. To understand a phenomenon is to be able to predict the outcome of a laboratory or controlled environment experiment where "other" variables are held constant and stochastic processes are limited. With the single exception of heavenly motion, physical scientists predict outcomes in controlled environments. They are no better, for example, at predicting the real world's meteorological phenomena than economists are at predicting the real world's economic events.

All etic approaches may in practice then, if not in principle, be nonpredictive. Behaviorism, for example, may in some cases at least predict behavior when its psychologists control environmental contingencies. In an uncontrolled setting, however, in which, as Skinner (1953) has acknowledged, no two environments are alike and one might add are ever static, the ability to predict would imply a knowledge that is inconceivable. Sociobiological analysis, as I have argued elsewhere (1980), is similarly bankrupt with respect to the relationship of its capacity to describe genetic factors and their likely influence on its capacity to predict. It is doubtful, moreover, that this situation will change sufficiently to permit Wilson's expectation (1978, p. 215) to materialize that biologically grounded social sciences will "mature into predictive disciplines."

All theoretical approaches to behavior may in fact be inherently nonpredictive. This is true to the extent that the focus of control resides within the individual, or indeed within the organism more generally. The well-known Harvard or Chicago law of animal behavior (it seems to have independently evolved in a number of places) states that "under precisely controlled conditions, an animal does as it damn well pleases" (Wald 1965, p. 40). It was the "essential unpredictability of animal behavior," Wald writes, that led to the formulation of the law.

Self-Direction

What Harré and Secord (1972, p. 8) refer to as self-direction—"the person acting as an agent directing his own behavior"—is equivalent to Sperry's concept (1965, 1980) of self-determination. In contrast to free will, self-determinism assumes that all behavior is caused, but in contrast to the standard doctrine on determinism, it also assumes that conscious choice is itself a causal determinant of our behavior—without denying the fact of other determinants, whether genetic or environmental in nature (see chapter 2).

Clearly a theory of behavior that acknowledges self-direction or self-determinism presupposes an emic plan of study—applied to individuals as well as to groups. In general, an emic approach becomes necessary to the extent that the individual or the group under study is considered to be characterized by self-direction. Let us briefly consider this proposition by reference to (1) different species, (2) individual development, and (3) individual differences.

1. Self-selection, self-direction, and animal life. Although all of animal life is characterized by self-selection, only man can be considered generally self-directed. Insect life, for example, is blindly purposive and therefore especially suitable for etic analysis along sociobiological lines. As awareness increases among higher forms of life, a wholly etic approach becomes inappropriate, as Jane Goodall's study of chimpanzees (1971) clearly indicates.

Only in humans, however, are self-selection and self-direction generally equivalent, and therefore an emic approach is especially called for. The very existence of self-selecting behavior nevertheless argues for genetic predispositions, while the failure to realize one's goals suggests genetic and environmental constraints. But over and above these influences, man himself directs the course of his own behavior. As Luckmann (1979, p. 60; emphasis in the original) writes, our species is uniquely characterized by the presence of a personal identity, which he defines as the *"central long-range control of its behavior by an individual organism."*

2. Ontogeny and self-direction. As children develop intellectually, they should become more capable of self-direction. In particular, when they reach Piaget's fourth and final stage of formal operations, wherein they become able to transcend their immediate environment in time and place, they also become able to direct their behavior toward the fulfillment of future goals.

An emic approach may nonetheless be in order well before this period is reached during early adolescence. Stoufe (1979, p. 836), for example, argues that "viewing children as active participants in their own experience is essential": "At least by the second half year, the infant's reaction to events is subjective; it is determined by evaluative processes within the infant, as well as by objective information." Kohen-Raz (1977, p. 104; emphasis in the original) writes that in the mental growth of the child the "activities *initiated* by the child himself seem to play a major role."

3. Individual differences and self-direction. As Crook (1980, p. 270) points out, people differ in the extent to which they believe the outcomes of their behavior are due to their own control and initiative or due to forces outside themselves. "Internalists" are those who believe the "focus of control" to lie within themselves; "externalists" believe it to lie elsewhere. Crook observes, based upon recent studies, that such subjective feelings appear to have their own objective impact; internalists are in fact more self-reliant, self-initiating, and independent in their actions. Thus what to a behaviorist is the myth of autonomy turns out for Crook to be "a crucial emergent in the realization of the self."

The internalist no doubt derives some of his feelings of autonomy from a favorable emotional climate, such as warm and supportive parents. In addition to such differences in emotional background, important cognitive differences are sure to exist. Based upon a middle-class American sample, Kohlberg and Gilligan (1971, p. 1065) conclude that "a large proportion of Americans never develop the capacity for abstract thought." Almost half of American adults, they state, "never reach adolescence in the cognitive sense" of actually attaining the Piagetian state of formal-operational reasoning.

It is plausible to speculate, as does Granit (1977, p. 82; cf. E. White 1969; 1972), that "normal levels of consciousness in a population" are "determined by a Gaussian distribution curve": "Some people seem to maintain a much higher level of conscious awareness than others." Conscious awareness, rational powers, and self-direction, it is plausible, are roughly related and normally distributed.

Mayr (1967, p. 836) has written, we recall, that "the human achievements—those things that distinguish man from the animals and are responsible for man's civilization, art, literature, and science—were achieved by less than one percent of the human population, by those in the upper tail of the curve of human variation in inventiveness, imagination, perseverance, and ability to think clearly." At the pinnacle of a normal curve for self-direction, we likely find individuals who are also influencing the course of the life of others in society. They are the leaders in various areas of life and thereby constitute a special focus of study within the sciences of human action.

To study self-directed and influential individuals, an emic approach is more than appropriate; it is essential. One cannot begin to understand a creative person without understanding the insights and goals that activate him. If this necessity means that the observer finds no basis for successfully predicting future action on the part of the creative person, he must realize that such a person himself has no sure basis for predicting his own future action. (Cf. "Fodor's First Law of the Nonexistence of Cognitive Science": The "more global a cognitive science is, the less anybody understands it" [Fodor 1983, p. 107].)

An emic plan of study brings us back to the concepts of Weber's *verstehen* (Weber 1947) and Cooley's "sympathy" (Cooley 1922). Both concepts enjoin the observer to recreate insofar as possible the interior life of the actor in order to understand the motivation and meaning of his external behavior. Such an approach characterized historical idealism dating to Dilthey and before (cf. Hughes 1958, ch. 6) and can be traced to the influence of Kant and Vico. We are reminded by Vico that man is uniquely

qualified to understand his social world—as against the physical world—
because he made it.

POLITICAL ACTION, EMIC ANALYSIS, AND ONGOING
POLITICAL BIOGRAPHY

> ... the first book can be, in the formal sense, nothing but a personal
> history which while written as a novel was to the best of the author's
> memory scrupulous to facts, and therefore a document; whereas the
> second, while dutiful to all newspaper accounts, eyewitness reports,
> and historic inductions available, while even obedient to a general style
> of historical writing, at least up to this point, while even pretending
> to be a history (on the basis of its introduction) is finally now to be
> disclosed as some sort of condensation of a collective novel—which is
> to admit that an explanation of the mystery of the events at the Pen-
> tagon cannot be developed by the methods of history—only by the
> instincts of the novelist. The reasons are several, but reduce to one.
> Forget that the journalistic information available from both sides is so
> incoherent, inaccurate, contradictory, malicious, even based on error
> that no accurate history is conceivable. More than one historian has
> found a way through chains of false fact. No, the difficulty is that the
> history is interior—no documents can give sufficient intimation: the
> novel must replace history at precisely that point where experience is
> sufficiently emotional, spiritual, psychical, moral, existential, or su-
> pernatural to expose the fact that the historian in pursuing the expe-
> rience would be obliged to quit the clearly demarcated limits of historic
> inquiry. So these limits are now relinquished. The collective novel
> which follows, while still written in the cloak of an historic style, and,
> therefore continuously attempting to be scrupulous to the welter of a
> hundred confusing and opposed facts, will now unashamedly enter that
> world of strange lights and intuitive speculation which is the novel.
> —Norman Mailer, *The Armies of the Night* (1968)

To understand political action, one must understand the political actors
who are responsible for it. This statement seems unexceptional, especially
in relation to leaders of the stature of Kemal Atatürk, Charles de Gaulle,
or Chou En-lai. How can their individual impact be denied—or understood
without directly examining the unique basic ideas and aims underlying their
political careers? Yet too often political explanations, in an attempt to
build a nomothetic science, discount individual and intellectual factors;
they in consequence favor an etic to the exclusion of an emic approach.

To see the unduly etic character of prevailing political explanation, let
us look first at the three models of international decision making—"the
rational actor," "organizational process," and "governmental politics"—
that are described by Allison (1971) in his discussion of the 1962 Cuban

missile crisis. I shall next assess a psychohistorical approach and then consider, finally, an emic approach to political action in the form of an "ongoing political biography."

The "rational actor" model in international politics in effect personalizes the national government, and as Allison (1971, ch. 1) describes it, views it as a single monolithic decision maker. Within this framework action rather than behavior is stressed; in other words, the purposes or intentions of the national entity provide the basis of explanation (Allison 1971, p. 10). In addition this model assumes rationality, for explanation consists not only in indicating what aim the government was attempting to achieve but also how specific actions were reasonable in the light of these aims. Despite its emphasis on rational action, this model is basically etic in its approach, as we shall see shortly.

The organizational process model replaces the national government as actor with a loose coalition of groups over which political leaders preside; and it is this constellation of groups each with its own aims and policies that provides the alternatives and, especially because of bureaucratic parochialism and inertia, the constraints for decision making by the leader (Allison 1971, pp. 79, 80, 87, 88). This analysis, as applied to the Cuban missile crisis, focuses upon its group determined characteristics that "set the drumbeat to which the actors marched" (Allison 1971, p. 102). As might be surmised, this approach too is largely etic.

The governmental politics framework substitutes for a unitary governmental actor, or even for a conglomerate of organizational actors, a wide variety of individual players in the political game, situated in a wide variety of organizational positions, and characterized by interests that reflect the influence of those positions and their own overriding concern in furthering their personal influence and advancement. Thus governmental actions are viewed as the resultant of the political "bargaining" for position and influence by a plurality of players, perhaps none of whom may have intended the actual outcome: the outcome results not as a considered solution to a problem but rather from the interplay of various unequal and conflicting group interests. Thus intentions and outcomes will seldom correspond, and rational action exists only in the world of the scholar and the idealist, not that of the practicing politician. This approach, which does acknowledge the importance of individual actors and which could well be placed, as will be indicated, within an overarching ethological or sociobiological perspective, also permits a greater emphasis to be placed upon an emic plan of study, as we shall see—but still insufficiently so.

Now does it make any difference which theoretical model one applies to the Cuban missile crisis? Of course it does. Allison (1971, p. 251) concludes: "Spectacles magnify one set of factors rather than another, and thus not only lead analysts to produce different explanations of problems ... but also influence the character of the analyst's puzzle, the evidence he assumes to be relevant, the concepts he uses in examining the evidence, and what he takes to be an explanation." Indeed so.

Allison's conclusion here is, for our purposes, twofold. First, it signifies that one's choice of a theoretical model matters. Theories have causal potency and thus consequences. Presumably such is true of ideas in general, of what Popper (1979, p. 229; emphasis in the original) has referred to as *"purposes, deliberations, plans, decisions, theories, intentions* and *values,"* and which, he has argued, can change the nature of the physical world.

Second, Allison's conclusion acknowledges the importance of theoretical diversity. When each theory is capable of making a difference in interpretation, it becomes important to identify the particular theoretical perspective of the scholar in discussing his conclusions. Theoretical diversity is found, moreover, within each of the models; each is far from monolithic in its individual formulation. Thus with respect to the rational actor model, Allison (1971, p. 13) observes that "in most respects, contrasts in the thinking of Morgenthau, Hoffmann, and Schelling, could not be more pointed."

Recognition of such diversity of veiwpoint need not, I should add, imply an inseparable relativism. It is possible to maintain that within any framework some observers are more sophisticated and sensitive than others and that especially the most gifted observers will escape the narrow confines of any conventional theory. One may indeed transcend the conventional formulations of one's own theory, as when Marx protested that he was not a Marxist. Allison is himself a good example of an observer transcending the theoretical models that he compares and contrasts, refusing to be conceptually straitjacketed and demonstrating the necessary capability not to be. The observer, to the extent that he transcends the theories that he compares, we might further note (cf. chapter 3), acts as an artist rather than as a scientist.

We might here remind ourselves of the point made by Almond and Genco (1977, p. 493) in the epigraph to this chapter: "Insofar as they [social scientists] acknowledge the importance of scientific memory, scientific creativity, calculative strategies, goal seeking, and problem solving in their own work, they must in some degree acknowledge these qualities in the human and social material they investigate and seek to explain." As Kissinger (1982) puts it succinctly in the foreword to his memoir, *Years of Upheaval,* "I alone am responsible for the contents of this book as I am for the actions it describes." Can one hope to understand Henry Kissinger (or Zbigniew Brzezinski) as scholars without understanding the intellectual content of their work? Do their intellectual ideas and aims cease to matter once they assume governmental roles? Are Kissinger and Brzezinski interchangeable either as scholars or as actors? What is critically deficient in the three models of international decision making that Allison considers (and in Allison's work generally) is a consideration of the impact that individual decision makers might have based at least in part on the intellectual understanding that they possess.

To provide just one example from the Cuban missile crisis: Nowhere in Allison's three models is there a place even for an extended treatment of the role of President John F. Kennedy, let alone of his intellectual background. Are such personal and intellectual considerations simply irrelevant? Certainly his role as president could not be dismissed as unimportant, but that role nonetheless could be viewed as defined by external forces. If so, John F. Kennedy, the man as distinct from the president, might be conveniently and safely ignored. But can he be? Certainly not if this admittedly sympathetic portrait by Schlesinger (1967, p. 113) has any validity at all to it:

It was autonomy which this humane and self-sufficient man seemed to embody. Kennedy simply could not be reduced to the usual complex of sociological generalizations. He was Irish, Catholic, New England, Harvard, Navy, Palm Beach, Democrat and so on; but no classification contained him. He had wrought an individuality which carried him beyond the definitions of class and race, region and religion. He was a free man, not just in the sense of the cold-war cliche, but in the sense that he was, as much as a man can be, self-determined and not the servant of forces outside him.

Here is an intellectual portrait from Sorenson (1966, p. 24):

He had a limitless curiosity about nearly everything—people, places, the past, the future. Those who had nothing to say made him impatient. He hated to bore or to be bored. But he enjoyed listening at length to anyone with new information or ideas on almost any subject, and he never forgot what he heard. He read constantly and rapidly—magazines, newspapers, biography and history (as well as fiction both good and bad). At times, on a plane or by a pool, he would read aloud to me a paragraph he found particularly forceful.

Is it possible that these qualities of mind might have had a bearing on the outcome of the Cuban missile crisis? In Robert F. Kennedy's account (1969, p. 127) we are told that

Barbara Tuchman's *The Guns of August* had made a great impression on the President. "I am not going to follow a course which will allow anyone to write a comparable book about this time, *The Missiles of October*," he said to me that Saturday night, October 26. "If anybody is around to write after this, they are going to understand that we made every effort to find peace and every effort to give our adversary room to move. I am not going to push the Russians an inch beyond what is necessary."

Writing twenty years following the crisis, Sidey (1982, p. 26) was to conclude: "In all the subsequent analysis of the Cuban crisis, scholars and participants have dwelt on nuclear balances, geography and diplomatic tactics. It just could be that Barbara Tuchman, author of the *Guns of August*, was as important as the U.S. Navy."

Let us turn once more to Allison's three models to see how such considerations are treated.

The Rational Actor Model

This model stresses the aims and intentions of the actor, but may in fact omit the actor, at least in the form of the actual living decision maker. Thus Allison (1971, p. 5) notes that "strategic analysts concentrate on the logic of action in the observer of an actor" and he comments (p. 247) that in contrast to the other two models, the first model "seems somewhat disembodied."

To the extent that actual persons are represented in its analysis, this model substitutes the intentions of the observer for those of the actor. Thus Allison (1971, p. 9) cites Schelling: "You can sit in your armchair and try to predict how people will behave by asking how you would behave if you had your wits about you. You get free of charge, a lot of vicarious, empirical behavior." Vicarious, yes; empirical, no. To substitute Schelling for Khrushchev or Brezhnev (as shorthand for the Soviet government) or whoever else is to get Schelling's thinking, not anyone else's. What if one substituted an organizational process or government process theorist for Schelling? Allison has suggested, we recall, how much difference it can make. It is just possible that the Soviet leader will not share the assumptions made in accord with a particular rational actor model (just as other scholars might not).

It is of course also clear that what might be rational for one Soviet leader might not be for another. An aging Brezhnev still alive and holding on to power in 1989 or a conservative Ligachev, ascending to power at that time, just might not, unlike Gorbachev, see it in the Russian national interest to withhold Soviet tanks from a restive Eastern Europe. And, on the other hand, none of them, if holding power in 1962, might have found it in the Soviet interest to place, like Khrushchev, missiles in Cuba. Indeed, until Gorbachev, according to Sestanovich (1990), the very concept "national interest" would have been eschewed by a Soviet leader in favor of "the advancement of the international class struggle."

To begin to understand the aims and intentions of another individual, it is necessary to recreate insofar as possible both the unique set of circumstances *and* the intellectual and personological qualities that appear to motivate that individual. That is the essence of an emic approach. *Verstehen*, which is its hallmark, means understanding the individual from his own point of view, not imposing upon him the frame of reference that is created by the observer. And of course an emic plan of study assumes actual individuals, not abstract entities such as "the American government." There is simply no substitute, as Hilsman (1981, p. 143) argued, for dealing with "the making of decisions by identifiable human beings."

The rational actor model is forced to minimize the individuality of the decision maker and hence the importance of individual differences because it accepts (as more or less do the other models discussed by Allison) the logic of explanation associated with Carl G. Hempel (Allison 1971, p. 278). Yet Hempel (1963, p. 358), in setting forth his covering law model of explanation, as pointed out in chapter 4, described a rational actor model that made a number of dubious assumptions: That there is a uniform rationality that is applicable across the board to any situation; and that all rational agents partake of this rationality and hence can be expected to act accordingly in any given circumstance. Hempel's concession that a rational agent may perform a given act "with high probability" rather than "invariably" only acknowledges the possibility that all manner of other factors impossible to predict may interfere with the original expectation.

Allison (1971, pp. 286–87) notes that theories are not to be discarded merely because they are "unrealistic." But as Almond and Genco (1977, p. 493) point out, explanatory schemes such as Hempel's that are based upon the hard physical science have "only a limited application to the social sciences" for they assume a uniformity and regularity of behavior that deny the "complexities of human and social reality." It is not merely counterfactual to deny a central role to conscious purposes and to the importance of individual difference; it is highly counterproductive. As Allison (1971, pp. 145, 174) observes in his description of the governmental politics model, people who share power differ about what should be done and "the differences matter": "The peculiar preferences and stands of individual players can have a significant effect on government action. Had someone other than Paul Nitze been head of the Policy Planning Staff in 1949, there is no reason to believe that there would have been an NSC–68. Had MacArthur not possessed certain preferences, power, and skills, U.S. troops might never have crossed the neck [in Korea]."

To leave out of one's theory the special qualities of the top decision makers is to leave out one of the major determinants of the decision; and it is done so needlessly in a vain attempt to achieve a systematic science of explanation and prediction. We have already seen that etic approaches (indeed any approach in the sciences of human action) cannot achieve *scientific* prediction (cf. Thurow 1977). Thus the would-be hard scientific approach attains neither understanding nor prediction, whereas at least the first objective may be attainable. For as one has declared, "Understanding is the enemy of prediction" (cited in Allison 1971, p. 272). Or perhaps equally: Prediction is the enemy of understanding.

Organizational Process Paradigm

This approach gains by replacing the mythical and monolithic actor of the state by "a constellation of loosely allied organizations"; but it too loses the individual within an etic framework, one that views decisions as largely determined by the constraints imposed by big impersonal organizations. Consequently, this approach does not escape the thrust of the critique directed toward the rational actor model.

Just as the first model calls attention to the importance of aims and intentions, however, the organizational process paradigm performs a service by focusing on the routines and inertia of large bureaucratic unity and the barriers and limits that they may place upon the once elegant designs of decision makers both within and without those units. It is essential nonetheless not also to lose sight of the individual decision maker, even within an organizational context.

Governmental Politics Paradigm

This model, in explicitly recognizing the importance of individual actors, makes a significant advance over the preceding two models. It nonetheless suffers two defects: It is insufficiently emic and cognitive in its approach.

The governmental politics paradigm is etic insofar as it assumes an individual to be under environmental (or genetic) direction. The models that Allison (1971, ch. 5) describes are basically environmentalist in their emphasis: "Positions define what players both may and must do," and "Where you stand depends on where you sit" (Allison 1971, pp. 165, 176). The latter aphorisms have both horizontal and vertical dimensions, which an ethologist or sociobiologist would be tempted to translate into territoriality and dominance relations and thereby to shift the approach from environmental to genetic direction in its emphasis. In either case, the resultant denial of self-direction yields an etic plan of study.

The governmental politics paradigm does acknowledge the differential skills of the various actors and their potentially critical impact upon any outcome. But the emphasis remains upon political bargaining and conflict, not upon ideas brought about by the different positions held by the participants and not upon any ideas and objectives that might motivate them. Allison (1971, p. 146) writes, "politicking lacks intellectual substance."

It may be true that memoirs dealing with political events tend to idealize and depoliticize them.[1] Thus Allison (1971, p. 146) observes that "both Sorenson and Schlesinger present the efforts of the ExCom in the Cuban missile crisis as essentially rational deliberation among a unified group of equals." But their memoirs may not err equally on both counts; that is, in revealing the divisions, clashes, and inequality in relationship among the participants, one does not necessarily show that rational and intellectual considerations were not at play. It is indeed within the context of this third model that Allison (1971, p. 218) does allude to the intellectual impact of Tuchman's *Guns of August* upon the president. Political decisions may in part be the resultant of competing pressures, but they may also be influenced by the aims and concerns of the leading participants.[2]

Now these three models do not necessarily have to be taken as mutually exclusive approaches; they can also be viewed as complementary, each highlighting forces that the others ignore. Indeed, as I have argued, all three models remain inadequate in considering the conscious aims of the individual participants. Insofar as human action is multifaceted and demands such recognition, more than a single perspective is required. Ultimately the study of human action is the study of a rich and unique multilayered complexity that denies any simple ordering along a single dimension. This situation, in the study of foreign affairs, leads Allison (1971, p. 272) to acknowledge

that the best explanations of foreign affairs are insightful, personalistic, and non-cumulative. . . . The field is so unstructured that each scholar is encouraged to make a personal contribution by expressing his understanding in a vocabulary that captures what is unique about his insight. Such insights, however, are not easily applied by less brilliant students of foreign policy to new cases. Consequently, perceptive analyses of particular happenings tend simply to illuminate these occurrences rather than to contribute to an accumulating body of systematic knowledge.

Allison appears reluctant to accept such a conclusion, but he may well have described the nature of human action as applied both to the realm of social science scholarship and to foreign policy making.

A Psychohistorical Approach

It remains for us to consider an approach that would appear directly to confront the question of the role of the individual in history: the psycho-historical. Mazlish (1976, p. 18) draws a rather sharp distinction between psychohistory that treats individuals and that which treats the group. The former, which Mazlish terms "life-history" and which constitutes our immediate concern, he defines as being "primarily concerned with the motives of an individual, suitably psychoanalyzed, of course, and the way in which these personal motives are shaped by the culture and society as well as by his genetic factors, etc."

Psycho- or life-history, then, is clearly concerned with human action; but insofar as individual motives are explained by internal or external factors beyond that individual's control, this approach is etic rather than emic. A Freudian psychoanalytic approach typically de-emphasizes self-direction. As Lifton (1979, p. 27; emphasis in the original), puts it, "Freud's fundamental discoveries—of the significance of man's individual and collective past—provide the basis for psycho-history. Yet on the other hand these same Freudian principles, when applied with closed-system finality, tend to reduce history to *nothing but recurrence* (or "repetition-compulsion") and thereby to eliminate virtually all that is innovative, or even accumulative in the story of man." The rigid Freudian psychohistorian tends to operate from a fixed, uniform framework and to impose it upon the individuals under study. But any elaborate theoretical framework for psychohistorical study carries with it the danger of prejudging its subject matter, of interpreting it only through the prism of its narrow conceptual focus. Thus Manuel (1971, p. 200), in referring to Erikson's eight-stage model of human development, writes that "the historian should be warned that the selection of materials to fill the boxes of the eight stages may make of the scheme a self-fulfilling prophecy."

Humanist psychologists, in their desire to provide a corrective to a static and somber Freudian outlook, risk the opposite error of imposing an overly dynamic and optimistic view on its subject matter. Consider, for example, Lifton's emphasis (1979, p. 3) upon the emergence of "Protean man," by which he means "a relatively new life style, characterized by interminable exploration and flux, by a self-process capable of relatively easy shifts in belief and identification." This admittedly post-Freudian construct may err as equally in the other direction, viewing an individual as endlessly dynamic and innovative. I have elsewhere (1979) argued that the Maslowian concept of the self-actualized individual, rather than being an empirically based one in fact imposes on any person so characterized a prior and uniform set of characteristics such as "spontaneity" and "a democratic character structure."

Who can deny that for some historical personages, Hitler and Nixon being only two notable recent examples, psychopathological tendencies have importantly affected their actions? No observer can simply dismiss irrational and unconscious forces; but neither can one dismiss the impact of ideas and objectives (and the skills necessary, with some success, to reach them that any imposing historical figure, indeed including a Hitler, must possess). A psychohistorical approach, in short, like any other approach cannot afford to be narrowly etic in nature.

ONGOING POLITICAL BIOGRAPHY

I now return to my key assumption that to understand political action we must understand the political actors responsible for it. As Somit (1981, p. 168; emphasis in the original) writes, "In the final analysis, political behavior is *human behavior*. Only human beings are political actors."

To this end, I have proposed an ongoing political biography (see E. White 1972). Political observers would begin to study political figures reaching certain positions of prominence—and either continue their study or have other budding scholars succeed them while their subjects-under-study remained in the political limelight. Political scientists in the 1950s, for example, would have begun a study of Richard Nixon upon his election as vice-president in 1952, Lyndon Johnson upon his becoming Senate minority leader in 1953, and John F. Kennedy after his receiving national prominence following the 1956 Democratic convention. Each year all such students of political biography might meet together with experts in relevant areas of public opinion research, substantive policy areas, and so forth, for a diagnosis of present and future policy.

The question is not, pejoratively, whether to indulge in political journalism, but whether the insights of the contemporary life and social sciences might not be brought to bear upon all the significant aspects of modern history. Political scientists may by default leave the ongoing analysis of individual decision making to political journalists (from whose example we might indeed profit), but to do so is needlessly to accept self-imposed limitations to our political understanding. There is no good reason why an individual response in a current poll should be considered scientific, but not the in-depth biographic treatment of current decision makers—or that the latter should become a scientifically respectable enterprise at the hands of professional historians a generation later.

Ongoing political biography is not at all intended to exclude, or to subsume, continuing work in such important areas of concern as public opinion, electoral analysis, and the like. Nor does it necessarily imply the end of the nomothetic study of political elites. Neither does it replace a more traditional emphasis on political institutions studied within their unique historical and social contexts (cf. Johnson 1989, ch. 3). To extend and to apply Masters's (1989) dictum—that social science must avoid "either/or" answers—both institutional and biographic approaches can be combined. Such a panclectic endeavor is one that I have attempted (forthcoming) in trying to understand the struggle over the ratification of the U.S. Constitution.

Or, as perhaps overstated by a perceptive observer of (if a marginal participant in) the second Reagan administration (Noonan 1990, p. 65): "I decided that history really is biography, and you could map the probable trajectory of an administration by making a deep study of the chief and his chiefs." Elsewhere (1986, 1990) I attempt just this (postdictively of course) in an effort to explain the course of events that led to Watergate during Nixon and Iran Contra during Reagan. These papers of mine at least attempt to provide a framework for future study along these lines.

Ongoing political biography does, on the other hand, argue the necessity for the in-depth analysis of significant political actors. Manley (1969) contends, for example, that the desire for generalization among academic congressional observers has invariably led to a neglect of the individual actor; yet he submits—and subsequently documents in an excellent case study—that one cannot understand the workings of the House Ways and Means Committee in the 1960s without understanding its chairman, Wilbur Mills. Neither, I submit, can one understand Chicago politics after 1955 without some understanding of Richard J. Daley, or of French politics after 1958 in the absence of Charles de Gaulle, or American foreign policy following 1969 without taking into account Henry Kissinger. Certainly the course of Russian and Eastern European affairs after 1985 cannot be understood without some consideration of Gorbachev (see Tatus 1991).

Ongoing political biography does not, however, imply a "great man" interpretation of politics. It merely acknowledges—to paraphrase Mayr (1963a)—that no two political leaders are alike and that heredity and environment make a contribution to nearly every politically relevant trait. It also fully acknowledges the role that ideas and objectives may play in political decisions. Public officials under study may indeed be inept as well as influential; but the raison d'être for studying them does not thereby cease; for the inept as well as the influential may independently affect policy. In the final analysis, ongoing political biography does little more than fully endorse and systematically extend Lasswell's long-ignored dictum (1930, p. 1): "Political Science without biography is a form of taxidermy."

There are, to be sure, serious problems in giving each political Johnson his academic Boswell. Let us briefly consider two of these: the problems of access and of foreknowledge. If easy access existed, a political scientist might well emulate what Smith, Bruner, and White attempted in their *Opinions and Personality* (1953) or what Cattell (1966) would attempt if public officials were legally required to submit to a program of testing (although of course going beyond such standardized and therefore etic testing). Such access is, needless to say, presently out of the question, especially throughout much of the world. Does this fact constitute grounds for bypassing the problem entirely? The prior question, however, is: Are political actors ever causal agents? If the answer is in the affirmative, then the student of politics has no scientific alternative but to consider his subject accordingly and to do whatever he can under the existing constraints.

The problem of foreknowledge inheres in the effort to discern the intentions of the political actor. For clearly there are times when he does his best to conceal those intentions from everyone except those closest to him; and at other times he himself may not know just what course of action he is going to pursue. In either case, the observer may hazard an intelligent guess, but he will hardly be engaging in scientific prediction.

Winston Churchill (cited in *Encyclopaedia Britannica*, 1962, S.V. "Free Will"), as was pointed out in chapter 2, described how the German ambassador to London was recalled for not having foreseen David Lloyd George's Mansion House speech: "How could he know what Mr. Lloyd George was going to do? Until a few hours before, his own colleagues did not know. Working with him in close association, I did not know. Until his mind was definitely made up, he did not know himself." Similarly, when Senator Jacob K. Javits of New York was deliberating whether or not to run for a fifth term in 1980, Javits himself (1981, p. 496) reports, "All the 'experts' had theories . . . and just as many of them reported I would run as reported I would not. It amused me to see those serious predictions and stories from 'sources' because I did not yet know the outcome."[3]

Even known intentions, finally, may become removed from the political arena, as through assassination (e.g., Sadat in 1982), political coup (e.g., Khrushchev in 1964), or resignation (e.g., de Gaulle in 1969). Yet there is clearly no certain way for the political scientist to predict any of these occurrences. True indeed. There are, one can only conclude, inherent limits to our political understanding; and these limits cannot be evaded simply by avoiding the factors that give rise to them.

Now to conduct ongoing political biography, I suggest observers follow a plan of study that is (1) empirical, (2) emic, and (3) panclectic. Let me discuss these three guidelines in turn.

Ongoing Political Biography That Is Empirical

An observer of a political leader should begin his or her study by being empirical in a descriptive sense. Contemporary ethology proves instructive in this respect. As Somit (1976b, pp. 314–15; emphasis in the original) observes: "Ethology has made great advances by looking at actual animal behavior. This may stimulate political scientists . . . to return to the study of *actual* political behavior. One of the great ironies of contemporary political science is that so many of those marching under the banner of 'behavioralism' have turned their backs upon behavior itself. Political science . . . will profit . . . if the ethological inspiration reorients us to the study of what people do, and how they behave in day-to-day political life."

A truly empirical approach will not be tied to a narrowly nomothetic science, one that Jane Goodall (1971) found inappropriate even for the study of chimpanzees; Goodall, one recalls, refused to deal with her subjects in a wholly impersonal, objective manner, but insisted on describing, even naming them with an eye to their individuality and character. The use of ethological description should prove liberating in allowing, indeed encouraging, the observer to make full use of all the data to be observed. Nonverbal communication, for example, has been typically ignored by the behavioralist, but Masters (1982, 1989) in his path-breaking studies points the way for political scientists to enrich their research with insights derived from such observation.

Empirical description should be ongoing as well, for the undeniable reason that the course of the life of each individual is marked by irreversible change—physically and psychologically. One need not fully endorse Lifton's "Protean man" to appreciate that political leaders are especially prone to changes in viewpoint as both they and circumstances change. Who in 1962 would have envisioned a decade later the visit of Richard Nixon, as president of the United States, to Mao's China? Who, in 1973, in the aftermath of the Yom Kippur war, would have guessed that within five years Anwar Sadat would be journeying to Begin's Jerusalem?[4] If leaders, like all of us, change, then observers must attempt to track this change continuously.

Ongoing Political Biography That Is Emic

The logical place to begin discovering an individual's aims and intentions is in his or her own description of them. As Allport (1960, p. 101) has put it: "When we set out to study a person's motives, we are seeking to find out what that person is trying to do in this life—including, of course, what he is trying to avoid and what he is trying to be. I see no reason why we should not start our investigation by asking him to tell us the answers as he sees them."

Such an emic approach was taken by Broder in his *Changing of the Guard* (1980), a study of the younger generation of new leaders in American social, economic, and political life as the United States entered the 1980s. Broder (1980, p. 467) states his aim and method as follows:

This book was conceived essentially as large-scale reporting project. It was an effort to draw a portrait of a generation posed on the brink of taking power. There were no obvious models for this particular kind of journalistic history, so I relied on the techniques used every day in reporting less ambitious topics.

What this has meant is that the portrait has been sketched largely in the words of the book's subjects. They were asked to describe the experiences that had propelled them into the political process and to discuss the attitudes and outlook they have toward themselves, their predecessor, their colleagues and rivals, their country and its government.

It was probably best to leave that task largely to them, for they, after all, are the ones who are going to be running this country.

The sensitive observer, in attempting to recreate the concrete circumstances and unique mentality that leads to a particular decision no doubt plays the role more of the artist than of the scientist. So be it. Indeed, he must take special care that he does not impute his own outlook or one that might be considered characteristic of a certain type of individual (e.g., the "American" or "bourgeoisie" mentality, or that of "economic man") to his subject. The question is not, after all, what I, the observer, or the "typical" Democrat or Republican, and so forth would do under certain conditions, but what the unique subject would do. It is this attempt to understand the subject's own special perspective that constitutes the basis of an emic approach grounded in *verstehen*.

Ongoing Political Biography That Is Panclectic

The need for a panclectic approach to the human personality is clear: It is evident, as Allport (1969) asserts, that no one theory can encompass all the facets of the personality. Moreover, taken to their extremes various theories prove contradictory and therefore all cannot be equally valid. The observer must be prepared to transcend a particular approach even if, as earlier suggested, such transcendence involves the sensitive judgment of the artist rather than the comprehensive system of the scientist. Selected with discretion, the use of different theories should prove complementary rather than exclusive.

Let me illustrate the need for panclecticism by referring to possible conflicts between the first two approaches to ongoing political biography: the empirical and the emic. An emic plan of study, as we have seen, assumes self-direction. It is indeed difficult to think of any major political leader who would not be considered self-directed. But it is nonetheless conceivable that for reasons of health, a leader might become mentally incapacitated (e.g., Brezhnev) and his basis for making decisions might shift to one less informed by conscious aims.[5]

It also has to be kept in mind, as suggested earlier, that the capacity for conscious thought and purpose will undoubtedly vary from individual to individual. Kissinger (1982, p. 1208) has commented:

If politics is the art of the possible, stature depends on going to the very limits of the possible. Great statesmen depend on going to the very limits of the possible. Great statesmen set themselves high goals yet assess unemotionally the quality of material, human and physical, with which they have to work; ordinary leaders are satisfied with removing frictions or embarrassments. Statesmen create; ordinary leaders consume. The ordinary leader is satisfied with ameliorating the environment, not transforming it; a statesman must be a visionary and an educator.

The observer must empirically determine whether he is dealing with a greater or lesser mortal and adjust his approach accordingly. Thus it is entirely fitting, indeed necessary, to treat the Founding Fathers, as has Diamond (1981), with an emphasis upon their political philosophy. Political ideas and long-term objectives were central to their efforts and lay entirely within their capabilities. An extended study of the political philosophy of Warren G. Harding would have less merit.

An emic approach also assumes that a person's stated goals can be taken at face value. To a surprising degree this may indeed be the case. Too many observers before World War II wrote off Hitler's *Mein Kampf* as mere propaganda. Boorstin (1953, pp. 76, 77) has written that "in this age of Marx and Freud we have begun to take it for granted that, if people talk about one thing, they must be thinking about something else. Ideas are treated as the apparatus of an intellectual sleight-of-hand, by which the speaker diverts the audience's attention to an irrelevant subject, while he does the real business unobserved." Yet historians like Bailyn (1967), for example, in studying the years preceding the American Revolution, have been convinced that many colonialists really did adhere to and were motivated by the principle incorporated by the slogan "No taxation without Representation" (see Reich 1988).

The empirical observer might nonetheless discern times when stated objectives and actual behavior diverge, as during Watergate. At such times an emic emphasis on aims and ideas might better be supplemented with one that acknowledges the impact of psychopathological deficiencies in character.

Thus there can be no blanket rule governing ongoing political biography, or the study of human affairs more generally. One can no more insist, Take everything at face value, than its reverse. The observer must use his or her own judgment in accord with the reality presented by the subject matter. In this respect, the sensitive observer and the superior statesman are similar: Each is guided by an overriding objective that is qualified in the light of the concrete circumstance.

NOTES

1. In his memoirs (*The Past Has Another Pattern* [New York: Norton, 1982], p. 309), George W. Ball, one of the key participants in the Cuban crisis, writes that we "argued out all available courses of action in an intellectual interchange that was the most objective I ever witnessed in government—or, for that matter, in the private sector." But what if Ball is right?

2. Similarly, political infighting may occur within an academic context, but it need not entirely "deintellectualize" its participants.

3. Cf. Jimmy Carter (1982, p. 37) on his selection of a vice-presidential running mate in 1976: "I deliberately withheld a decision until all the interviews and investigations were completed, and I must admit that I changed my mind three or four times. Several people claimed they knew in advance whom I had chosen, and went to the news reporters with their "inside" information. There was a flood of stories by journalists, each claiming to have found the one person who knew my secret choice! But no one could have known, because I did not know myself."

4. To be sure, Kissinger (1982) suggests (in retrospect) that Sadat's intentions to seek a rapprochement with Israel were already present even before 1973 and, paradoxically, were in large part responsible for the war.

5. Of course in such cases major decisions tend to be postponed or to be assumed by others, underlining the significance of the capacity for conscious choice in decision making.

7

The Restoration of Politics as the Master Art

If, then, there is some end of the things we do, which we desire for its own sake . . . clearly this must be the good and chief good. Will not knowledge of it, then, have a great influence on life? . . . It would seem to belong, to the most authoritative art and that which is most truly the master art. And politics appears to be of this nature; for it is this that ordains which of the sciences should be studied in a state, and which each class of citizens should learn and up to what point they should learn them.

—Aristotle, *Ethics*, bk. 1, ch. 2

Now in my own circles we were starting to recognise that in modern societies . . . it is politics that for good or ill proves to be the decisive shaper of society. . . . Especially important was the idea that government, far from simply reflecting class interests, had to be regarded as a feat of craftsmanship, something made by and therefore open to numerous variations. I learned—and for this I remain grateful to Hannah Arendt—that politics had to be scrutinized in its own right, and not just as an index of social conflicts.

—Irving Howe (1982)

I argue in this chapter for the renewed recognition of politics as "the master art," the Aristotelian insight that our most inclusive and sovereign ends and decisions are likely to be of a political character. Political choices will necessarily retain their sovereign character even in more libertarian regimes and may well acquire a more urgent status with the increasing capacity of man to direct the future course of his own evolution.

I first lay the evolutionary and neurobiological ground for this conclusion by discussing the larger implication of the roles played by purpose and rational self-awareness in human affairs. As one important consequence of the inevitability of human choice, I contend that the fact-value dichotomy widely held within the social sciences is in fact untenable. This argument, with its implicit return to the classical philosophy of natural right (cf. Strauss 1960), parallels the recent work of others, especially Masters (1981, 1983, 1989; see also Arnhart 1982).

HUMAN PURPOSE AND REASON: THE UNTENABILITY OF THE FACT-VALUE DICHOTOMY

"Even lower animals move where the conditions please them," Dobzhansky (1970b, p. 93) has written. In chapter 5 I called attention to the work of Darlington (1969), Dobzhansky (1970b), and Mayr (1970, 1982) in which it is asserted that animals generally seek congenial environments.

All animals, Aristotle noted long before Darwin, have the "capacity for pleasure and pain" and hence "the desire for what is pleasant" (*De Anima* II, 2. iii, 2–5).

Four general observations follow: First, modern biology, in restoring purpose to the world of life, restores it to the universe. Everyday experience and scientific theory are hereby joined, since it is clear from our own lives that our acts have no meaning aside from the purposes that appear to impel them.

In rejecting the earlier mechanism of Darwinian evolution, contemporary life scientists have not adopted a supernaturalistic vitalism. Rather purpose is restricted to the living organism, which is therefore neither controlled by external vitalistic forces nor reduced internally to the sum of the particles—the physical-chemical matter—that compose it.

Nevertheless, if life is a natural part of the universe and if purpose is a natural part of life, then purpose is itself a natural part of the universe. As Jonas (1966, p. 233) has written, the existential view of man as "having been flung into an indifferent universe" ought rather to be expressed by saying that "life-conscious, caring, knowing feeling has been 'thrown up' by nature." Whatever the purpose of purpose, it appears to exist by nature and therefore in nature. This view may not correspond to the entirely teleological cosmos described by the ancients, but it is closer to it than to an existentialist chaos or mechanistic order of things.

Second, purpose presupposes an actor and, further, an actor with a nature of its own. For the alternative implies an organism either passively controlled by external forces (as in a mechanistic natural selection, or behaviorist psychology) or somehow creating its own nature in the process of living (as in existentialist thought). Against this latter possibility, Camus (1958, p. 16) has written: "Analysis of rebellion leads at least to the suspicion that, contrary to the postulates of contemporary thought, a human nature does exist, as the Greeks believed." The existentialist alternative, as Jonas (1966, p. 228) points out, takes man out of nature and makes him "a law unto himself," yet leaving it still problematic what it is in the universe that can alienate man and what it is in man that can be alienated. As against either an environmental determinism or existentialist freedom, modern biology postulates the presence in each organism of a genotype—its complete genetic endowment. While not producing fixed traits in higher animals, the genotype does set natural limits and establish natural tendencies in behavior.

Purposeful activity is thus defined within the context of the natural endowment of the organism. Clearly mice and men pursue different ends, at least most of the time. Such is the case even for members of the same species and at least in part for genetically based reasons. Goldstein (1963) had called attention to our "preferred behavior" patterns, that is, our natural avoidance of environments that we do not find congenial and our natural preference for those that we do. These preferences or aversions, even if environmentally influenced, are ultimately rooted in the biological organism; they are, as Grene puts it, "*the best expression of this organism's essential nature*"—leaving Goldstein's position open to the "accusation" of "essentialism," in which case, Grene (1969, p. 41; emphasis in the original) notes, his critics "will be quite right."

Third, purpose implies potentiality. Inborn capacities must exist prior to their fulfillment—or to the desire or effort to fulfill them. Purposeful activity therefore presupposes a distinction between things as they are at a given moment and things as they might be or are desired to be in the future—a distinction emphasized in Aristotle. This distinction is crucial for understanding the fallacious use of the conventional objection to a naturalistic ethics—the naturalistic fallacy.

As propounded by G. E. Moore (1959) in explicit reference to a naturalistic ethics based on evolution, it argues against the view, to use Alexander Pope's phrase, that "Whatever is, is right." Thus it argues against simply taking the evolutionary process as such as the arbiter of our ethical choices. For as Moore (1959, p. 42) writes, "If everything natural is equally good, then certainly ethics, as it is ordinarily understood, disappears; for nothing is more certain, from an ethical point of view, than some things are bad and others are good; the object of Ethics is indeed, in chief part, to give you general rules whereby you may avoid the one and secure the other." But Moore does not go further; he does not deal with the Aristotelian concept of potentiality and the distinction between what is and what can be, the latter nonetheless derived ultimately from the former. As Simpson (1969, p. 132; emphasis in the original) has written: "It is undoubtedly illogical to conclude that what *is* therefore *ought to be*. It is, however, equally illogical to make that the basis for a further conclusion that decision as to what ought to be cannot be based on consideration of what is—in other words, that naturalistic ethics are impossible."

Moore discusses health as an illustration of the naturalistic fallacy. A naturalistic ethic would say, according to Moore (1959, p. 42), that "Nature has fixed what health shall be: and health, it may be said, is obviously good; hence in this case nature has decided the matter." But what, asks Moore, is this natural definition of health? "We may presume that what is meant is that it is normal; and that when we are told to pursue health as a natural end, what is implied is that the normal must be good." But is this really so, asks Moore (1959, p. 43)? "Was the excellence of Socrates or of Shakespeare normal? It is, I think, obvious in the first place, that not all that is good is normal; that, on the contrary, the abnormal is often better than the normal: peculiar excellence, as well as peculiar viciousness, must obviously be not normal but abnormal." Moore does not deny that health may be good, but rather he denies that the normal is necessarily good—that assertion being precisely an example of the naturalistic fallacy.

But why does Moore insist on defining health as normal? As he himself acknowledges (1959, p. 42; emphasis in the original) in the course of his definition: "I can only conceive that health should be defined in natural terms as the *normal* state of an organism: for undoubtedly disease is a natural product." But precisely because of disease, not to mention accidents and criminal assaults, we may readily observe a wide disparity in the conditions of health around us; that is, ranging from bad to excellent. Normally, perhaps, most of us endure a cold or two and a few odd ailments a season. But some individuals remain "fit as a fiddle" from one day to the next, while still others at the opposite extreme are chronically ill. Both groupings are of course few in proportion to the larger population; and that is just the point. We do not judge the excellence of eyesight on the basis of how well most people (unaided) can in fact see. They may well be born near-sighted. We judge on the basis of the potential excellence of human sight, as exemplified by those exceptional individuals whom nature has endowed with it.

Moore (1959, p. 43) does note that when we use the word *healthy*, "we do not mean the same thing by it as the thing which is meant in medical science." But that, it turns out, is because by it "we do commonly imply good"—as if doctors did not share with their patients that common evaluation. Moore, however, does not in any further way elaborate on this distinction in the definition of health. If he had, he might have concluded that most people do take a simplistic view, that is, equating health with normality; whereas in medical science a more sophisticated standard prevails—although carrying the matter still further, one would find that even most doctors have been criticized, as by the eminent biochemist and nutritionist Roger Williams (1973), for not distinguishing more fully between normal and excellent health. Ultimately, in other words, the definition of these terms and hence the standards they imply are properly set by the most excellent and hence exceptional practitioners in medicine. And to discuss these critical issues *only* on the basis of ordinary language usage is to fall prey precisely to the naturalistic fallacy.

We may recall that Moore notes that disease as well as health is "natural." But his discussion of ethics nonetheless proceeded from the perspective of man. Conceivably one might be concerned that the interests of germs be adequately represented; but if one really wanted to be genuinely egalitarian one would permit the germs to figure out their own system of ethics. In discussing the naturalistic ethics of Julian Huxley, Simpson (1969, p. 142) observes: "An ethic so derived is not based on the nature of evolution but on the nature of man. It is the better for that."

If we choose to treat the nature of man—something that could be and was done with insight before Darwin—we must then distinguish man from the other animals. According to Dobzhansky, (1970b, p. 25) animal life varies in relationship to its environment from narrow specialization to conscious mastery; the "really successful species are masters of a variety of environments," with the human species "perhaps most adept at this strategy." Only man can "choose, control, and to some extent create environments."

Simpson (1971, p. 284), in terming man "almost the only animal that really exerts any significant degree of control over the environment," declares:

Even when viewed within the framework of the animal kingdom and judged by criteria of progress applicable to that kingdom as a whole and not peculiar to man, man is thus the highest animal. It has often been remarked (perhaps again merely "*pour épater les bourgeois*") that if, say, a fish were a student of evolution it would laugh at such pretensions on the part of an animal that is so clumsy in the water and that lacks such features of perfection as gills or a homocercal caudal fin. I suspect that the fish's reaction would be, instead, to marvel that there are men who question the fact that man is the highest animal. It is not beside the point to add that the "fish" that made such judgments would have to be a man!

Three important consequences flow from man's unique rational mastery of the environment. First, there is itself the scientific recognition of consciousness as an independently significant agent. Here common sense, tradition, and science join in what for the twentieth century is a truly revolutionary development. As against either the materialist reductionism of the molecular biologist or the environmental determinism of the behaviorist psychologist, we have argued that only an emergent mentalist approach accounts for the known facts of the operation of the brain. Only a knowledge of the overall interdependent workings of the entire brain can begin to explain human consciousness—which dominates the activity of any given part of the brain rather than vice versa. The understanding of an organ, as Aristotle had realized, depends not on its "material elements, but (on its) composition, and the totality of the form, independently of which (it has) no existence" (*De Partibus Animalium* I, 5, 35–37).

Second, the fact that man is by nature endowed with purpose and reason in a world fraught with dangers and brimming with possibilities means that conscious choice is inescapable. Accordingly, distinctions between good and bad choices—involving what Aristotle referred to as deliberation (*Ethics* 3, iii)—are also inevitable; and hence modern biologists, such as Simpson, Huxley, and Waddington have all described man as being uniquely the ethical animal. Infants and mental defectives alone escape ethical choices.

It follows, finally, that the fact-value distinction fails to hold in any pure form. For if we are necessarily purposive and rational beings, these qualities inform our subsequent thoughts and actions—quite apart from the question of their wisdom or morality. Any fact that we describe as a scientist, for example, we also value because it fits into a larger theory that we also value. All facts are in the end "value-laden" since they must in the end derive their selection and their significance from the values or purposes that infuse our lives.

Sperry (1981, p. 8; emphasis in the original) provides a neurobiological basis for the preceding contention: "The classic *fact-value* and *is-ought* dichotomies of philosophy logically dissolve in the context of cerebral processing. The operations of the brain are already by nature richly replete with established values and value determinisms, both inherent and acquired, with the result that incoming facts regularly interact with and shape values." With the capacity for choice, then, comes the inescapability of choice—choice that has consequences and that is necessarily value-laden.

THE RESTORATION OF POLITICS AS THE MASTER ART

Values, then, are an integral and inherent part of our lives. In an emergent model of the brain, moreover, "subjective phenomena including values," writes Sperry (1981, p. 7), "are brought into the causal sequence in human decision-making and behavior generally." Thus they enjoy a certain autonomy. Wilson (1978, p. 5), on the other hand, would "biologize" ethics and politics: "science may soon be in a position to investigate the very origin and meaning of human values, from which all ethical pronouncements and much of political practice flow." Yet Wilson (1978, p. 196; cf. Lumsden and Wilson 1983, p. 174) also asserts that the deterministic hold of our evolutionary past "is not so tight that it cannot be broken by an exercise of will" and that "a biology of ethics" will "make possible the selection of a more deeply understood and enduring code of moral values." But how else are an "exercise of will" and a "more deeply understood" ethics possible save in the light of an active and autonomous human self-awareness?

Now precisely these properties, we recall, characterize Popper's view (Popper and Eccles 1977, p. 120) of the human mind: "the mind is, as Plato said, the pilot. It is not, as David Hume and William James suggested, the sum total, or the bundle, or the steam of its experiences: this suggests passivity. . . . Like a pilot, it observes and takes action at the same time." Ethical choices must, from this view, enjoy more than a merely epiphenomenal status. Furthermore, they may themselves constitute prime examples of "downward causation," that is, they may directly influence the course of our lives. Political choices are especially of this character. Such choices, which as Almond and Genco (1977) point out constitute the "center" of "political reality," usually involve decisions binding on entire populations of individuals. They inescapably affect the lives of us all.

Of the models of the brain treated in the first two chapters, only an emergent model grants a causal role for politics. In the others—whether molecular, genetic, physiological, or environmental in nature—political values and choices become reduced to some lower set of explanation factors. Politics taken on its own terms disappears. Only an emergent view, to repeat, fully acknowledges the causal potency of political ideas and decisions (cf. Corning 1983).

Needless to say, Aristotle long ago recognized the status of politics as "the master art." And it remains an art because no ironclad set of general principles can be set down to apply to all circumstances. Rather the political leader, like an experienced doctor, must be able to administer general laws within particular circumstances (cf. Aristotle *Ethics* 1181b).

The centrality if not the primacy of political philosophy, together with the inapplicability of the fact-value distinction to the political realm, has been forcefully argued by Masters (1989). In my review of his book (1991) I raised the question, therefore, directly with him as to whether his analysis in fact implies the return to politics of Aristotle's designation of it as the "master art." Masters's thoughtful reply (1991) anticipates both the resistance to this claim from other disciplines as well as the imposing responsibility that such a claim would entail: Scholars "never like to be told that their own discipline or 'field' is lower in status than some other field (professors are, like howler monkeys, territorial animals). Still, one can make a strong case for Aristotle's articulation of the sciences, and therewith his claim of a special status for political science. This entails, however, a special responsibility: to have a 'master' or supervisory status over other sciences and arts. We are so far from this today that it is probably best to be more modest about our discipline."

One emerging area of decision making in any event that will have the most momentous consequences involves the course of our own evolution. "The human species can change its own nature," Wilson (1978, p. 208) states. "What will it choose?" What indeed? Whatever the choice, the possibility of making it reminds us, in Dawkins's words (1977, p. 63) of the "evolutionary trend toward the emancipation of survival machines as executive decision-takers from their ultimate masters, the genes." With our brains we can control the genes that have programmed those brains.

Political choices so staggering in their implications would surely justify the idea of politics as the master art. But political choices, after all, have always had significant consequences. So what I have termed "the restoration of politics as the master art" may therefore be a misnomer. We should more properly refer to the restoration of the recognition of human politics as the master art and to the challenge that that restoration of recognition may imply for the future of the field.

Bibliography

Adelson, Joseph. "Adolescence and the Generation Gap." *Psychology Today* 12 (1979): 33–37.

Alexander, Richard. *Darwinism and Human Affairs*. Seattle: University of Washington Press, 1979.

Alliluyeva, Svetlana. *Only One Year*. New York: Harper Colophon, 1969.

Allison, G. *Essence of Decision*. Boston: Little, Brown and Co., 1971.

Allport, Gordon. *Becoming*. New Haven: Yale University Press, 1955.

———. *Personality and Social Encounter*. Boston: Beacon Press, 1960.

———. *The Person in Psychology*. Boston: Beacon Press, 1969.

Almond, G. A. *A Discipline Divided*. Newbury Park, Calif.: Sage, 1990.

Almond, Gabriel, and Stephen Genco. "Clouds, Clocks and the Study of Politics." *World Politics* 29 (1977): 489–523.

Anastasi, Anne. "Heredity, Environment, and the Question 'How'?" In *Individual Differences*, ed. Anne Anastasi, 170–86. New York: John Wiley, 1967.

Antonovics, Janis. "The Effects of a Heterogeneous Environment on the Genetics of Natural Populations." *American Scientist* 59 (September-October 1971): 593–99.

Ardrey, Robert. *The Social Contract*. New York: Atheneum, 1970.

Arnhart, Larry. "Charles Darwin and the Declaration of Independence." Paper prepared for delivery at the Annual Meeting of the American Political Science Association, Denver, Colorado, September 1982.

Arterton, C. F. "The Impact of Watergate on Children's Attitudes Toward Political Authority." *Political Science Quarterly* 89 (1974): 259–88.

Ayala, F. J., and Theodosius Dobzhansky, eds. *Studies in the Philosophy of Biology*. Berkeley and Los Angeles: University of California Press, 1974.

Ayala, F. J., and J. Valentine. *Evolving*. Menlo Park, Calif.: Benjamin/Cummings, 1979.

Baars, B. S. *The Cognitive Revolution in Psychology*. New York: Guilford, 1986.

Bailey, K., and P. Gravning. "Physiological Growth, IQ, and the Development of Political Awareness in Pre-Adolescence: A Longitudinal Test of the Maturation/Environment Hypothesis." Paper prepared for delivery at the Annual Meeting of the Southwestern Political Science Association, Fort Worth, Texas, 1979.

Bailyn, B. *The Ideological Origins of the American Revolution.* Cambridge: Harvard University Press, 1967.

Baldwin, J. M. "Organic Selection." *Nature*, April 15, 1897.

Barash, David P. *Sociobiology and Behavior.* New York: Elsevier, 1977.

———. *The Whisperings Within.* New York: Harper and Row, 1979.

Beck, Aaron T. "Cognitive Therapy of Depression." In *Depression*, ed. Paula Clayton. New York: Raven Press, 1983, 117–149.

———. "Cognitive Therapy: A 30-year Retrospective," *American Psychologist* 46:368–375; April, 1991.

Beckner, M. O. "Vitalism." In *The Encyclopedia of Philosophy*. Vol. 8, 253–56. New York: Macmillan, 1968.

Bell, Daniel. "The Year 2000—The Trajectory of an Idea." *Daedalus* (Summer 1967): 639–51.

———. *The Social Sciences Since the Second World War.* New Brunswick, N.J.: Transaction Books, 1982.

Berelson, Bernard, and Gary Steiner. *Human Behavior.* New York: Harcourt, Brace and World, 1964.

Berger, Peter, and T. Luckmann. *The Social Construction of Reality.* Garden City, N.J.: Doubleday, 1967.

Bergson, Henri. *Creative Evolution.* New York: Henry Holt, 1971.

Berlin, I. *Vico and Herder.* New York: Viking Press, 1976.

Bernstein, Richard. *The Restructuring of Social and Political Theses.* New York: Harcourt Brace Jovanovich, 1976.

Birch, Charles. "Chance, Necessity and Purpose." In *Studies in the Philosophy of Biology*, ed. F. J. Ayala and Theodosius Dobzhansky, 225–40. Berkeley and Los Angeles: University of California Press, 1974.

Bisiach, Edoardo. "The (Haunted) Brain and Consciousness." In *Consciousness in Modern Science*, ed. A. J. Marcel and E. Bisiach, 101–20. Oxford: Clarendon, 1988.

Boddy, J. *Brain Systems and Psychological Concepts.* New York: John Wiley, 1978.

Boehm, Christopher. "Rational Preselection from Hamadryas to *Homo Sapiens*: The Place of Decisions in Adaptive Process." *American Anthropologist* 80 (1978): 265–96.

Bohm, David. "Some Remarks on the Notion of Order." In *Towards a Theoretical Biology*, Vol. 2, ed. Conrad Hal Waddington, 18–41. Chicago: Aldine, 1968.

Bohr, Niels. *Atomic Physics and Human Knowledge.* New York: John Wiley, 1963.

Boorstin, Daniel S. *The Genius of American Politics.* Chicago: University of Chicago Press, 1953.

Boring, Edwin. *The Physical Dimension of Consciousness.* New York: The Century Co., 1933.

Bovet, D. "Nature, Nurture and the Psychological Approach to Learning." In *Brain and Human Behavior*, ed. H. G. Kurczman and John C. Eccles, 324–40. New York: Springer-Verlag, 1972.

Brandon, R. N. "Adaptation and Evolutionary Theory." *Studies in the History and Philosophy of Science* 9 (1978): 181–206.

Bridgman, P. W. *The Way Things Are*. Cambridge: Harvard University Press, 1959.

Broder, David S. *Changing of the Guard*. New York: Simon and Schuster, 1980.

Bronowski, Jacob. *Nature and Knowledge: The Philosophy of Contemporary Science*. Eugene: Oregon System of Higher Education, 1969.

Brzezinski, Z. *The Grand Failure*. New York: Scribner's, 1988.

Burns, J. M. *Leadership*. New York: Harper and Row, 1978.

Campbell, D. T. "Downward Causation in Hierarchically Organized Biological Systems." In *Studies in the Philosophy of Biology*, ed. F. J. Ayala and T. Dobzhansky, 179–86. Berkeley and Los Angeles: University of California Press, 1974.

Campbell, J. H. "The New Gene and Its Evolution." In *Rates of Evolution*, ed. K.S.W. Campbell and M. F. Day, 283–309. London: Allen & Unwin, 1987.

Campbell, Ruth. "Cognitive Neuropsychology." In *Growth Points in Cognition*, ed. G. Claxton. London: Routledge, 1988, 109–137.

Camus, Albert. *The Rebel*. New York: Vintage, 1958.

Carlyle, Thomas. "On Heroes, Hero-Worship, and the Heroic in History." In *The Varieties of History*, ed. Fritz Stern. New York: Meridian Books, 1956, 23–48.

Carter, Jimmy. *Keeping the Faith*. New York: Bantam Books, 1982.

Caton, Hiram. "Domesticating Nature: Thoughts on the Ethology of Modern Politics." In *Sociobiology and Human Politics*, ed. Elliott White, 99–133. Lexington, Mass.: Lexington Books, 1981.

Cattell, Raymond B. *The Scientific Analysis of Personality*. Chicago: Aldine, 1966.

———. *Abilities: Their Structures, Growth, and Action*. Boston: Houghton Mifflin, 1971.

Childs, Barton. "Commentary II." In *Developmental Human Behavior Genetics*, ed. K. Warner Schaie et al., 119. Lexington, Mass.: Lexington Books, 1975.

Chomsky, Noam. "The Case Against B. F. Skinner." *New York Review of Books* 17 (December 30, 1971), 18–25.

Churchland, Patricia Smith. "Reduction and the Neurobiological Basis of Consciousness." In *Consciousness in Contemporary Science*, ed. A. Marcel and E. Bisiach, 273–304. Oxford: Clarendon Press, 1988.

Clark, Austen. *Psychological Models and Neural Mechanisms*. Oxford: Clarendon Press, 1980.

Claxton, Guy, ed. *Growth Points in Cognition*. London: Routledge, 1988.

Cocconi, G. "The Role of Complexity in Nature." In *The Evolution of Particle Physics*, ed. M. Conversi, 81–87. New York: Academic Press, 1970.

Cooley, C. H. *Human Nature and the Social Order*. New York: Charles Scribner's Sons, 1922.

Coopersmith, Stanley. *Antecedents of Self-Esteem*. San Francisco: W. H. Freeman, 1967.

Corning, Peter. "The Biological Bases of Behavior and Some Implications for Political Science." *World Politics* 23 (1971): 312–70.

———. "What Is Natural Selection?" Unpublished manuscript, 1981.

———. *The Synergism Hypothesis*. New York: McGraw-Hill, 1982.

———. *Politics and the Evolutionary Process*. New York: Harper and Row, 1983.

Cox, Catherine. *Genetic Studies of Genius*. Vol. 2, *The Early Mental Traits of Three Hundred Geniuses*. Stanford, Calif.: Stanford University Press, 1926.

Crick, Francis. *Of Molecules and Men*. Seattle: University of Washington Press, 1966.

———. "Thinking About the Brain." *Scientific American* 241 (1979), 219–33.

Crook, J. *The Evolution of Human Consciousness*. Oxford: Clarendon Press, 1980.

Crow, James F. "Mechanism and Trends in Human Evolution." In *Evolution and Man's Progress*, ed. H. Hougland and R. W. Burhoe. *Daedalus* (Summer 1961): 416–31.

Csikszentmihalyi, M. *Flow: The Psychology of Optimal Experience*. New York: Harper and Row, 1990.

Cutler, Neal. "Political Socialization Research as Generational Analysis: The Cohort Approach versus the Lineage Approach." In *Handbook of Political Socialization*, ed. S. A. Renshon, 294–328. New York: Free Press, 1977.

Darlington, Cyril Dean. *Genetics and Man*. New York: Schocken Books, 1969.

———. *The Evolution of Man and Society*. New York: Simon and Schuster, 1971.

Darwin, Charles. *On the Origin of Species by Means of Natural Selection*. London: Murray, 1859.

Davies, J. C. "Ions of Emotion and Political Behavior." In *Biology and Politics*, ed. Albert Somit, 97–126. Paris: Mouton, 1976.

———. "Political Socialization: From Womb to Childhood." In *Handbook of Political Socialization*, ed. S. A. Renshon, 142–71. New York: Free Press, 1977.

Davis, Bernard, and Patricia Flaherty, eds. *Human Diversity: Its Causes and Social Significance*. Cambridge, Mass.: Ballinger, 1976.

Dawkins, Richard. *The Selfish Gene*. New York: Oxford University Press, 1976.

———. *The Extended Phenotype*. San Francisco: W. H. Freeman, 1982.

Dawson, P. S. "A Conflict Between Darwinism Fitness and Population Fitness in *Tribolium* 'Competition' Experiments." *Genetics* 62 (1969): 413–19.

Dawson, R. E., and K. Prewitt. *Political Socialization*. Boston: Little, Brown and Co., 1969.

Delgrado, J.M.R. "New Trends in Behavioral Neurochemistry." In *Behavioral Neurochemistry*, ed. J.M.R. Delgrado and F. V. de Feudis, 1–10. New York: Spectrum, 1977.

Delli Carpini, M. X. "Age and History: Generations and Sociopolitical Change." In *Political Learning in Adulthood*, ed. R. Sigel. Chicago: University of Chicago Press, 1989, 211–243.

Dember, William. "Motivation and the Cognitive Revolution." *American Psychologist* 29 (1974): 161–68.

Dennett, D. C. *Elbow Room*. Cambridge, Mass.: MIT Press, 1984.

Deutsch, Karl W. "On Political Theory and Political Action." *American Political Science Review* 63 (June 1971): 442–64.

Diamond, Martin. *The Founding of the American Republic*. Peacock, Mich.: Peacock Publishing, 1981.

Dillon, Lawrence. *The Inconstant Gene*. New York: Plenum Press, 1983.

Dimond, S. J. *Introducing Neuropsychology*. Springfield, Ill.: Charles C. Thomas, 1978.

Dingle, John H. "The Ills of Man." *Scientific American* 229 (September 1973), 76–89.

Dobzhansky, Theodosius. *Genetics and the Origin of Species*. New York: Columbia University Press, 1951.

———. *Heredity and the Nature of Man*. New York: Signet Books, 1966.

———. *The Biology of Ultimate Concern*. New York: New American Library, 1967a.

———. "Of Flies and Men." *American Psychologist* 26 (1967b): 41–48.

———. "On Types, Genotypes, and the Genetic Diversity in Populations." In *Genetic Diversity and Human Behavior*, ed. James N. Spubler, 1–19. New York: Viking Fund, 1967c.

———. *Genetics of the Evolutionary Process*. New York: Columbia University Press, 1970a.

———. *Mankind Evolving*. New York: Bantam Books, 1970b.

———. "Determinism and Indeterminism in Biological Evolution." In *Man and Nature: Philosophical Issues in Biology*, ed. Ronald Munson, 188–212. New York: Dell Books, 1970c.

———. "Chance and Creativity in Evolution." In *Studies in the Philosophy of Biology*, ed. F. J. Ayala and Theodosius Dobzhansky, 307–38. Berkeley and Los Angeles: University of California Press, 1974.

Dobzhansky, Theodosius, et al., eds. *Evolution*. San Francisco: W. H. Freeman, 1977.

Douglass, Frederick. *Narrative*. New York: Anchor Books, 1963.

Driesch, Hans. *The Science and Philosophy of the Organism*. Vols. 1 and 2. London: Adam and Charles Black, 1908.

Dubos, Rene. "Science and Man's Nature." *Daedalus* (Winter 1965): 223–34.

———. *A God Within*. New York: Charles Scribner's Sons, 1972.

Dunn, John. "Practicing History and Social Science on 'Realist' Assumptions." In *Action and Interpretation*, ed. C. Hookway and P. Pettit, 145–77. Cambridge: Cambridge University Press, 1977.

Easton, D. *The Political System*. New York: Alfred A. Knopf, 1953.

Easton, D., and R. Hess. "The Children's Political World." *Midwest Journal of Political Science* 6 (1962): 229–46.

Eccles, John C. *Facing Reality*. New York: Springer-Verlag, 1970.

———. *The Understanding of the Brain*. New York: McGraw-Hill, 1973.

———. "Cerebral Activity and Consciousness." In *Studies in the Philosophy of Biology*, ed. F. J. Ayala and T. Dobzhansky, 87–108. Berkeley and Los Angeles: University of California Press, 1974.

———. "Brain and Free Will." In *Consciousness and the Brain*, ed. G. C. Globus, G. Maxwell, and I. Savodnik, 101–24. New York: Plenum Press, 1976a.

———. "How Dogmatic Can Materialism Be?" In *Consciousness and the Brain*, ed. G. C. Globus, G. Maxwell, and I. Savodnik, 155–62. New York: Plenum Press, 1976b.

———. *The Human Mystery*. New York: Springer-Verlag, 1978.

———. *The Understanding of the Brain*. Rev. ed. New York: McGraw-Hill, 1979.

———. *The Human Psyche*. Berlin: Springer International, 1980.

Eccles, John C. *Evolution of the Brain*. London: Routledge, Chapman & Hall, 1989.

Eccles, J. C., and D. N. Robinson. *The Wonder of Being Human: Our Brain and Our Mind*. New York: The Free Press, 1984.

Eckberg, B., and L. Hill. "The Paradigm Concept and Sociology." *American Sociological Review* 44 (1979): 925–37.

Eckland, Bruce. "Genetics and Sociology: A Reconsideration." *American Sociological Review* 32 (April 1967): 173–94.

Edelman, G., and V. B. Mountcastle. *The Mindful Brain*. Cambridge, Mass.: MIT Press, 1978.

Edelman, G. *Neural Darwinism: The Theory of Neuronal Group Selection*. New York: Basic Books, 1987.

Edelman, Murray. *The Symbolic Uses of Politics*. Urbana: University of Illinois Press, 1964.

Edelman, Murray. *Constructing the Political Spectacle*. Chicago: University of Chicago Press, 1988.

Einstein, Albert, and Leopold Infeld. *The Evolution of Physics*. New York: Simon and Schuster, 1938.

Eisenstadt, S. N. *From Generation to Generation*. Chicago: Free Press, 1956.

———. "Archetypal Patterns of Youth." In *The Challenge of Youth*, ed. Erik Erikson, 29–50. New York: Anchor Books, 1965.

Elsasser, Walter. *Atom and Organism: A New Approach to Theoretical Biology*. Princeton: Princeton University Press, 1966.

Escalona, S. R. *The Roots of Individuality*. Chicago: Aldine, 1968.

Etkin, William. "Social Behavior and the Evolution of Man's Mental Faculties." *American Naturalist* 88 (May-June 1954): 129–42.

———, ed. *Social Behavior from Fish to Man*. Chicago: University of Chicago Press, 1967.

———. "A Biological Critique of Sociobiology." In *Sociobiology and Human Politics*, ed. Elliott White, 45–97. Lexington, Mass.: Lexington Books, 1981.

———. "Evolution of the Human Mind and Emergence of Tribal Culture: A Mentalist Approach." *Perspectives in Biology and Medicine* 28, no. 4 (1985): 498–525.

Eulau, Heinz. *The Behavioral Persuasion in Politics*. New York: Random House, 1963.

———. "Political Behavior." In *International Encyclopedia of the Social Sciences*. Vol. 12, 203–14. New York: Macmillan Co. and the Free Press, 1968.

Eysenck, Hans Jurgen. *The Biological Basis of Personality*. Springfield, Ill.: Charles C. Thomas, 1967.

Ferris, T. *Coming of Age in the Milky Way*. New York: Morrow, 1988.

Feuer, Lewis. *The Conflict of Generations*. New York: Basic Books, 1969.

Fodor, J. A. *The Modularity of Mind*. Cambridge, Mass.: MIT Press, 1983.

Frankel, Charles. "Sociobiology and Its Critics." *Commentary* 68 (July 1979): 43–47.

Freedman, Daniel G. "A Biological View of Man's Social Behavior." In *Social Behavior from Fish to Man*, ed. William Etkin, 152–88. Chicago: University of Chicago Press, 1971.

———. *Human Infancy: An Evolutionary Perspective*. Hillsdale, Ill.: Lawrence Erlbaum, 1974.

Freeman, Derek. "Comments on Papers by Leyhausen and Masters." In *Human*

Ethology, ed. M. von Cranach et al., 285–92. Cambridge: Cambridge University Press, 1979.

Freud, Sigmund. *Totem and Taboo*. New York: Free Press, 1931.

Fuller, John L. "Genes, Brains and Behavior." In *Sociobiology and Human Nature*, ed. Michael S. Gregory, Anita Silvers, and Diane Sutch, 98–115. San Francisco: Jossey-Bass, 1978.

Fuller, John L., and W. Robert Thompson. *Behavior Genetics*. New York: John Wiley, 1960.

Gallistel, C. R. "Animal Cognition." *Annual Review of Psychology* 40 (1989): 155–89.

Gardner, Howard. *The Mind's New Science*. New York: Basic Books, 1985.

Gaylin, William. "We Have the Awful Knowledge to Make Exact Copies of Human Beings." *New York Times Magazine*, March 5, 1972.

Gazzaniga, Michael. *The Social Brain*. New York: Basic Books, 1985.

———. *Mind Matters*. Boston: Houghton Mifflin, 1988.

Gelenberg, Alan J., and Gerald L. Klerman. "Drug Therapy in Long-term Treatment of Depression." In *Controversy in Psychiatry*, ed. J. P. Brady and H.K.H. Brodie, 281–301. Philadelphia: W. B. Saunders, 1978.

Glass, Bentley. "Science: Endless Horizons or Golden Age?" *Science* 171 (January 8, 1971): 23–39.

Gleich, J. *Chaos*. New York: Penguin, 1989.

Goldstein, K. *Human Nature in the Light of Psychopathology*. New York: Schocken Books, 1963.

Goodall, Jane. *In the Shadow of Man*. New York: Dell Books, 1971.

Gould, Stephen Jay. "Darwin's Retreat: Review of Peter James Vorzimmer's *Charles Darwin: The Years of Controversy*." *Science* 172 (1971): 677–78.

———. "Is a New and General Theory of Evolution Emerging?" *Paleobiology* 6 (1980a): 119–30.

———. "Sociobiology and the Theory of Natural Selection." In *Sociobiology: Beyond Nature/Nurture?*, ed. G. W. Barlow and J. Silverberg, 257–73. Boulder, Colo.: Westview Press, 1980b.

———. "But Not Wright Enough: Reply to Orzack." *Paleobiology* 7 (1981): 131–34.

———. *Wonderful Life*. New York: Norton, 1989.

Gould, Stephen Jay, and Richard C. Lewontin. "The Spandrels of San Marco and the Panglossian Paradigm: A Critique of the Adaptationist Programme." In *Proceedings of the Royal Society of London* 8205 (1979): 581–98.

Gould, Stephen Jay, and E. S. Verba. "Exaptation—A Missing Term in the Science of Form." *Paleobiology* 8 (1982): 4–15.

Granit, Ragnar. *The Purposive Brain*. Cambridge, Mass.: MIT Press, 1977.

Gregory, Michael S., Anita Silvers, and Diane Sutch, eds. *Sociobiology and Human Nature*. San Francisco: Jossey-Bass, 1978.

Grene, Marjorie. *Approaches to a Philosophical Biology*. New York: Basic Books, 1969.

———. "Sociobiology and the Human Mind." In *Sociobiology and Human Nature*, ed. Michael S. Gregory, Anita Silvers, and Diane Sutch, 213–24. San Francisco: Jossey-Bass, 1978.

Griffin, Donald E. *The Question of Animal Awareness*. New York: Rockefeller University Press, 1976.

———. "Humanistic Aspects of Ethology." In *Sociobiology and Human Nature*, ed. Michael S. Gregory, Anita Silvers, and Diane Sutch, 240–59. San Francisco: Jossey-Bass, 1978.

———. *Animal Thinking*. Cambridge: Harvard University Press, 1984.

Gunnell, John. "Deduction, Explanation and Social Scientific Inquiry." *American Political Science Review* 63 (1969): 1233–46.

———. *Political Theory: Tradition and Interpretation*. Englewood Cliffs, N.J.: Winthrop, 1979.

———. *Between Philosophy and Politics*. Amherst: University of Massachusetts Press, 1986.

Hagstrom, Warren P. *The Scientific Community*. New York: Basic Books, 1965.

Hamilton, William D. "The Genetical Evolution of Social Behavior. I, II." *Journal of Theoretical Biology* 7 (1964): 116, 17–51. Reprinted in *Group Selection*, ed. George C. Williams, 23–89. Chicago: Aldine, 1971.

Hampshire, S. *Thought and Action*. London: Chatto and Windus, 1959.

Handler, Philip, ed. *Biology and the Future of Man*. New York: Oxford University Press, 1970.

Hardy, Alister. *The Living Stream: Evolution and Man*. New York: Meridian. 1965.

Harré, R., and P. F. Secord. *The Explanation of Social Behavior*. Totowa, N.J.: Rowman and Littlefield, 1972.

Harris, Marvin. *The Rise of Anthropological Theory*. New York: Crowell, 1968.

Harth, Erich. *Windows on the Mind: Reflections on the Physical Basis of Consciousness*. New York: William Morrow and Co., 1982.

Harvey, S. K., and T. G. Harvey. "Adolescent Political Outlooks: The Effects of Intelligence as an Independent Variable." *Midwest Journal of Political Science* 14 (1970): 565–95.

Hawking, Stephen W. *A Brief History of Time*. New York: Bantam Books, 1988.

Heilbroner, R. *The World's Philosophers*. New York: Simon and Schuster, 1953.

Heisenberg, Werner. *Physics and Philosophy: The Revolution in Modern Science*. New York: Harper and Row, 1958.

Hempel, Carl. "Reasons and Covering Laws in Historical Explanation." In *Philosophy and History*, ed. Sidney Hook, 339–61. New York: New York University Press, 1963.

———. *Aspects of Scientific Explanation*. New York: Free Press, 1968.

Hempel, Carl, and P. Oppenheim. "The Logic of Explanation." In *Readings in the Philosophy of Science*, ed. H. Feigl and M. Brodbeck, 319–52. New York: Appleton-Century-Crofts, 1953.

Hess, Robert, and Judith Tormey. *The Development of Political Attitudes in Children*. Chicago: Aldine, 1967.

Hilgard, Ernest. "Consciousness in Contemporary Psychology." *Annual Review of Psychology* 3 (1980): 1–26.

Hilsman, Roger. *Strategic Intelligence and National Decisions*. Westport, Conn.: Greenwood, 1981.

Hines, Samuel, Jr. "Evolutionary Epistemology and Political Knowledge." In *Through the Looking Glass: Epistemology and the Conduct of Political In-*

quiry, ed. Marie Falco, 329–65. Washington, D.C.: University Press of America, 1979.

———. "Biopolitics and the Evolution of Inquiry in Political Science." *Politics and the Life Sciences* 1 (1982): 5–17.

Hirsch, Jerry. "Behavior Genetics and Individuality Understood." *Science* (1963): 136–42. Reprinted in *Heredity and Achievement*, ed. Daniel N. Robinson, 353–78. New York: Oxford University Press, 1970.

———. "Intellectual Functioning and the Dimension of Human Variation." In *Genetic Diversity and Human Behavior*, ed. James N. Spuhler, 19–33. New York: Viking Fund, 1967.

Holton, Gerald. "The Roots of Complementarity." *The Making of Modern Science: Biographical Studies. Daedalus* (Fall 1970): 1015–55.

Hook, Sidney. *The Hero in History*. Boston: Beacon Press, 1955.

Howe, Irving. *A Margin of Hope*. New York: Harcourt Brace Jovanovich, 1982.

Hubel, David. "The Brain." *Scientific American* 241 (1979), 44–53.

Hughes, David L. "Sociobiology: Another New Synthesis." In *Sociobiology: Beyond Nature/Nurture?*, ed. G. W. Barlow and J. Silverberg, 77–97. Boulder, Colo.: Westview Press, 1980.

Hughes, Stuart. *Consciousness and Society*. New York: Random House, 1958.

Hull, David. *The Philosophy of Biological Science*. Englewood Cliffs, N.J.: Prentice Hall, 1980.

Hunt, Earl. "Cognitive Science." *Annual Review of Psychology* 40 (1989): 603–23.

Huntington, Ellsworth. *Mainsprings of Civilization*. New York: Mentor Books, 1964.

Huxley, Julian. "The Open Bill's Open Bill: A Telenomic Enquiry." *Zoological Systematics* 88 (1960): 9–30.

Illich, Ivan. *Medical Nemesis: The Expropriation of Health*. New York: Pantheon, 1976.

Intellectuals and Change. Daedalus 101 (Summer 1972).

Intellectuals and Tradition. Daedalus 101 (Spring 1972).

Itzkoff, Seymour W. *The Form of Man: The Evolutionary Origins of Human Intelligence*. Ashfield, Mass.: Paideia, 1983.

———. *Triumph of Intelligence: The Creation of Homo Species*. Ashfield, Mass.: Paideia, 1985.

———. *Why Humans Vary in Intelligence: The Evolution of Human Intelligence*. Ashfield, Mass.: Paideia, 1987.

Jackson, J. G., and J. H. Jackson. *Infant Culture*. New York: Doubleday, 1978.

James, William. "The Sentiment of Rationality." In *Essays in Pragmatism*, ed. Alburey Castel. New York: Hafner, 1948.

———. *Psychology: Briefer Course*. New York: Collier Books, 1969.

Janis, Irving. *Victims of Group Think*. Boston: Houghton Mifflin, 1972.

Jaros, Dean. "Biochemical Desocialization: Depressants and Political Behavior." Paper presented at the Annual Meeting of the American Political Science Association, Los Angeles, California, 1970.

Javits, Jacob K. *Javits*. Boston: Houghton Mifflin, 1981.

Jaynes, Julian. *The Origin of Consciousness in and Breakdown of the Bicameral Mind*. Boston: Houghton Mifflin, 1976.

Jensen, Arthur. "How Much Can We Boost IQ and Scholastic Achievement?" *Harvard Educational Review* 39 (Winter 1969): 1–123.

Jerisold, A., J. Brook, and P. Brook. *The Psychology of Adolescence*. New York: Macmillan, 1978.

Jerison, Harry. *Evolution of the Brain and Intelligence*. New York: Academic Press, 1973.

Johanson, Donald C., and M. Eden. *Lucy*. New York: Simon and Schuster, 1981.

Johnson, N. *The Limits of Political Science*. Oxford: Clarendon Press, 1989.

Jonas, Hans. *The Phenomenon of Life*. New York: Harper and Row, 1966.

Kagan, Jerome. *The Nature of the Child*. New York: Basic Books, 1984.

Kaplan, Abraham. *The Conduct of Inquiry*. San Francisco: Chandler, 1964.

Karczmar, A. G., and John C. Eccles. *Brain and Human Behavior*. Berlin: Springer-Verlag, 1972.

Katz, Jack. *The Seductions of Crime: The Moral and Sensual Attractions of Doing Evil*. New York: Basic Books, 1988.

Kendall, W., and Wolfgang Panofsky. "The Structure of the Proton and the Neutron." *Scientific American* (June 1971), 60–77.

Kennedy, E. "The Looming 80's." *New York Times Magazine* 68, December 2, 1978.

Kennedy, R. *Thirteen Days*. New York: W. W. Norton, 1969.

Kety, Seymour. "Disorders of the Human Brain." *Scientific American* 241 (1979), 202, 217.

Kissinger, Henry A. *Years of Upheaval*. Boston: Little, Brown and Co., 1982.

Kitcher, Phillip, *Vaulting Ambition*. Cambridge, Mass.: MIT Press, 1985.

Klausmeir, H. J., et al. *Conceptual Learning and Development*. New York: Academic Press, 1974.

Klivington, Kenneth A. *The Science of Mind*. Cambridge, Mass.: MIT Press, 1989.

Koch, Helen L. *Twins and Twin Relations*. Chicago: University of Chicago Press, 1966.

Koestler, Arthur. *The Ghost in the Machine*. New York: Macmillan, 1967.

Kohen-Raz, R. *Psychological Aspects of Cognitive Growth*. New York: Academic Press, 1977.

Kohlberg, Lawrence, and Carol Gilligan. "The Adolescent as a Philosopher: The Discovery of the Self in a Postconventional World." *Daedalus* (Fall 1971): 1051–86.

Kolodin, Irving. *In Quest of Music*. New York: Doubleday, 1980.

Kreitler, H., and S. Kreitler. *Cognitive Orientation and Behavior*. New York: Springer-Verlag, 1976.

Kroeber, A. *Anthropology*. Rev. ed. New York: Harcourt, Brace, 1948.

Kuhn, Thomas. *The Structure of Scientific Revolutions*. Chicago: University of Chicago Press, 1962.

———. "Reflections on My Critics." In *Criticism and the Growth of Knowledge*, ed. Imre Lakatos and A. Musgrave. Cambridge: Cambridge University Press, 1970.

Kurczman, H. G., and John C. Eccles. "Introduction." In *Brain and Human Behavior*, ed. H. G. Kurczman and John C. Eccles, 1–22. New York: Springer-Verlag, 1972a.

———, eds. *Brain and Human Behavior*. New York: Springer-Verlag, 1972a.

Landes, David L., and Charles Tilly, eds. *History as Social Science*. Englewood Cliffs, N.J.: Prentice-Hall, 1971.

Langton, K. P. *Political Socialization*. New York: Oxford University Press, 1969.

Lasswell, Harold D. *Psychopathology and Politics*. Chicago: University of Chicago Press, 1930.

Lazarus, Arnold. *The Practice of Multimodal Therapy*. Baltimore: Johns Hopkins University Press, 1990.

LeDoux, J. E., and W. Hirst. *Mind and Brain*. Cambridge: Cambridge University Press, 1986.

Lenneberg, Eric. "The Neurology of Language." *Daedalus* 102 (Summer 1973): 115–34.

Lewontin, Richard C. *The Genetic Basis of Evolutionary Change*. New York: Columbia University Press, 1974.

———. "Adaptations." In *Scientific American, Evolution*. San Francisco: W. H. Freeman, 1978.

———. "Sociobiology as an Adaptationist Program." *Behavioral Science* 24 (1979): 5–14.

———. "Darwin's Real Revolution." *New York Review of Books*, June 16, 1983, 21–27.

Lieberman, David A. "Behaviorism and the Mind: A (Limited) Call for a Return to Introspection." *American Psychologist* 74 (1979): 319–33.

Lifton, Robert S. "On Psychohistory." In *Explorations in Psychohistory*, ed. Robert S. Lifton and S. Olson. New York: Simon and Schuster, 1979, 3–41.

Lipset, S. M. "Politics and Societies in the USSR." *PS* 23 (1990): 20–28.

Losco, Joseph. "Ultimate vs. Proximate Explanation: Explanatory Modes in Sociobiology and the Social Sciences." *Journal of Social and Biological Structures* 4 (1981): 329–46.

———. "Evolution, Consciousness and Political Thought." Paper presented to the Annual Meeting of the American Political Science Association, Denver, Colorado, 1982.

Luckmann, T. "Personality as an Evolutionary and Historical Problem." In *Human Ethology*, ed. Maria von Cranach et al. Cambridge: Cambridge University Press, 1979, 77–99.

Lumsden, Charles, and Edward O. Wilson. *Genes, Mind, and Culture*. Cambridge: Harvard University Press, 1981.

———. *Promethean Fire*. Cambridge: Harvard University Press, 1983.

Luria, S. E., and Zella Luria. "The Soluble and the Insoluble or Are Two Cultures Better Than One?" *Daedalus* (Winter 1975): 273–77.

Lykken, David T. "Research with Twins: The Concept of Emergenesis." *Psychophysiology* 19 (1982): 361–73.

MacDonald, K. B. *Social and Personality Development*. New York: Plenum, 1988.

MacLean, Paul D. *A Triune Concept of the Brain and Behavior*. Toronto: University of Toronto Press, 1973.

MacRae, Duncan. *The Social Function of Social Science*. New Haven: Yale University Press, 1976.

Malcolm X. *The Autobiography of Malcolm X*. New York: Grove Press, 1965.

Mandelbaum, M. *Purpose and Necessity in Social Theory*. Baltimore: Johns Hopkins University Press, 1987.

Manley, John F. "Wilbur Mills: A Study in Congressional Influence." *American Political Science Review* 63 (June 1969): 442–64.

Mannheim, Karl. "The Problem of Generations." In *Essays on the Sociology of Knowledge*, ed. Paul Kecskemiti, 276–321. New York: Oxford University Press, 1932.

Manuel, F. E. "The Use and Abuse of Psychology in History." *Daedalus* 100 (1971): 187–213.

Marcel, A. J., and E. Bisiach, eds. *Consciousness in Contemporary Science*. Oxford: Clarendon Press, 1988.

Margolis, Howard. *Patterns, Thinking, and Cognition*. Chicago: University of Chicago Press, 1987.

Maslow, Abraham H. *Motivation and Personality*. New York: Harper and Row, 1970.

Masters, Roger D. "Politics as a Biological Phenomenon." *Social Science Information* 14 (1975): 7–63.

———. "The Value—and Limits—of Sociobiology." In *Sociobiology and Human Politics*, ed. Elliott White, 135–65. Lexington, Mass.: Lexington Books, 1981.

———. "Nice Guys *Don't* Finish Last: Aggressive and Appeasement Gestures in Media Images of Politicians." Paper presented at the Annual Meeting of the American Association for the Advancement of Science, Washington, D.C., 1982.

———. "The Biological Nature of the State." *World Politics* 35 (1983): 181–93.

———. *The Nature of Politics*. New Haven: Yale University Press, 1989.

———. "Answers to Elliott White's Questions about *The Nature of Politics*." *Politics and the Life Sciences* 10 (1991):103.

Maxwell, Grover. "Comment." In *Consciousness and the Brain*, ed. G. C. Globus, G. Maxwell, and I. Savodnik, 329–58. New York: Plenum Press, 1976.

Maynard Smith, John. "Group Selection and Kin Selection." *Nature* 201 (1964): 1145–47.

———. *The Theory of Evolution*. Baltimore: Penguin Books, 1966.

Mayr, Ernst. *Animal Species and Evolution*. Cambridge: Harvard University Press, 1963a.

———. "Cause and Effect in Biology." In *Toward a Theoretical Biology*, ed. Conrad Hal Waddington, 42–54. Chicago: Aldine, 1963b.

———. "Biological Man and the Year 2000." In *Toward the Year 2000: Work in Progress. Daedalus* (Summer 1967): 832–36.

———. *Populations, Species, and Evolution*. Cambridge, Mass.: Belknap Press, 1970.

———. "The Nature of the Darwinian Revolution." *Science* 176 (June 2, 1972): 981–89.

———. *Evolution and the Diversity of Life*. Cambridge: Harvard University Press, 1976.

———. "Evolution." *Scientific American* 239 (September 1978), 47–55.

———. *The Growth of Biological Thought: Diversity, Evolution, and Inheritance*. Cambridge, Mass.: Belknap Press, 1982.

———. "Speciational Evolution or Punctuated Equilibria." *Journal of Social and Biological Structures* 12 (1989): 137–58.

Mayr, Ernst, and W. B. Provine, eds. *The Evolutionary Synthesis: Perspectives on the Unification of Biology*. Cambridge: Harvard University Press, 1980.

Mazlish, H. B. "What Is Psycho-history?" In *Variations of Psycho-history*, ed. G. M. Kren and L. H. Rappoport. New York: Springer-Verlag, 1976.

McDermott, Walsh. "Evaluating the Physician and His Technology." *Daedalus* (Winter 1977): 125–34.

McGuiness, D., and Karl Pribram. "The Neuropsychology of Attention: Emotional and Motivational Controls." In *The Brain and Psychology*, ed. M. C. Wittrock, 95–141. New York: Academic Press, 1980.

McKinnell, R. G. *Clonging: Nuclear Transplantation in Amphibia*. Minneapolis: University of Minnesota Press, 1978.

Mead, M. *Culture and Commitment: A Study of the Generation Gap*. Rev. ed. New York: Doubleday, 1978.

Merelman, R. "The Development of Political Ideology: A Framework for the Analysis of Political Socialization." *American Political Science Review* 63 (1969): 750–67.

Midgley, M. *Beast and Man*. Ithaca, N.Y.: Cornell University Press, 1978.

Miles, Ian. *The Poverty of Prediction*. Westmead, England: Saxon House/Lexington, 1975.

Milgram, Stanley. *Obedience to Authority*. New York: Harper and Row, 1974.

Miller, Eugene. "Historicism, Positivism, and Political Inquiry." *American Political Science Review* 66 (September 1972): 796–874.

Mills, S., and J. Beatty. "The Propensity Interpretation of Fitness." *Philosophy of Science* 46 (1979): 263–86.

Mishan, E. J. *Cost-Benefit Analysis*. London: George Allen and Unwin, 1971.

Mondimore, Francis M. *Depression, the Mood Disease*. Baltimore: Johns Hopkins University Press, 1990.

Monod, Jacques. *Chance and Necessity*. New York: Vintage, 1972.

Monroe, K., et al. "The Nature of Contemporary Political Science." *PS* 23 (1990): 34–43.

Moore, G. E. *Principia Ethica*. Cambridge: Cambridge University Press, 1959.

Moore, Mark. "What Sort of Ideas Become Public Ideas?" In *The Power of Public Ideas*, ed. R. Reich, 55–83. Cambridge, Mass.: Ballinger, 1988.

Morgan, C. H. *Habit and Instinct*. London: Arnold, 1896.

Mountcastle, V. B. "An Organizing Principle for Cerebral Function; the Unit Module and the Distributed System." In *The Mindful Brain*, ed. G. Edelman and V. B. Mountcastle, 7–50. Cambridge, Mass.: MIT Press, 1978.

Nagel, Ernest. "The Logic of Historical Analysis." In *The Philosophy of History in Our Time*, ed. Hans Meyerhoff, 203–18. Garden City, N.Y.: Doubleday, 1959.

———. *The Structure of Science*. New York: Harcourt, Brace and World, 1961.

Natsoulas, Thomas. "Perhaps the Most Difficult Problem Faced by Behaviorism." *Behaviorism* 11 (1983): 1–26.

Nie, N. H., S. Verba, and J. R. Petrocik. *The Changing American Voter*. Cambridge: Harvard University Press, 1976.

Noonan, P. *What I Saw at the Revolution*. New York: Random, 1990.

Olby, Robert. "Francis Crick, DNA, and the Central Dogma." In *The Making of Modern Science. Daedalus* 99 (Fall 1970): 938–79.

Omen, Gilbert. "Genetic Mechanisms in Human Behavioral Development." In *Developmental Human Behavior Genetics*, ed. K. Warner Schaie et al., 93–111. Lexington, Mass.: Lexington Books, 1975.

Oppenheimer, J. Robert. *Science and the Common Understanding*. New York: Simon and Schuster, 1954.

Ornstein, Robert. *The Psychology of Consciousness*. New York: Viking, 1973.

Orzack, S. H. "The Modern Synthesis Is Partly Wright." *Paleobiology* 7 (1981): 128–31.

Ospovat, D. *The Development of Darwin's Theory*. Cambridge: Cambridge University Press, 1981.

Parsons, Talcott. *Action Theory and the Human Condition*. New York: Free Press, 1978.

Pelletier, K. P. *Toward a Science of Consciousness*. New York: Delacorte, 1978.

Penfield, Wilder. *The Mystery of the Mind*. Princeton: Princeton University Press, 1975.

Peterson, Steven A. "The Human Brain and Hypostatizing." Paper presented at the Meeting of the International Society for Political Psychology, Washington, D.C., 1982.

Pfeiffer, John E. *The Emergence of Man*. New York: Harper and Row, 1969.

Piaget, Jean. *Insight, and Illusions of Philosophy*. New York: World, 1965.

———. *Structuralism*. New York: Basic Books, 1970.

———. "Operational Structures of the Intelligence and Organic Controls." In *Brain and Human Behavior*, ed. H. G. Kurczman and John C. Eccles, 273–398. New York: Springer-Verlag, 1972.

Pike, K. L. *Language in Relation to a Unified Theory of the Structure of Human Behavior*. The Hague: Mouton, 1967.

Plomin, Robert. "The Role of Inheritance in Behavior." *Science* 248 (1990): 183–88.

Polanyi, Michael. *Personal Knowledge*. London: Routledge, 1958.

Popper, Karl. *The Poverty of Historicism*. New York: Harper Torchbooks, 1961.

———. "Of Clouds and Clocks," 1965. Reprinted as chapter 6 of Karl Popper, *Objective Knowledge*. Oxford: Clarendon Press, 1979.

———. *Objective Knowledge*. Oxford: Clarendon Press, 1979.

Popper, Karl, and John C. Eccles. *The Self and Its Brain*. Berlin: Springer-Verlag, 1977.

Presthus, Robert. *Men at the Top*. New York: Oxford University Press, 1964.

Pribram, Karl. *Language of the Brain*. Englewood Cliffs, N.J.: Prentice-Hall, 1971.

———. "Problems Concerning the Structure of Consciousness." In *Consciousness and the Brain*, ed. G. C. Globus, G. Maxwell, and I. Savodnik, 297–314. New York: Plenum Press, 1976a.

———. "Self-Consciousness and Intentionality." In *Consciousness and Self-Reduction*, ed. G. E. Schwartz and D. Shapiro. New York: Plenum Press, 1976b, 109–130.

———. "Holographic Memory." *Psychology Today* 12 (1979), 70–86.

———. "The Brain as the Locus of Cognitive Controls in Action." In *Cognition in Human Motivation*, ed. G. d'Ydewalle and W. Lens. Louvain, Belgium: Leuven University Press, 1981, 215–253.

Price, Don K. "The Established Dissenter." In *Science and Culture. Daedalus* (Winter 1965): 84–117.

Prout, Timothy. "The Estimation of Fitness from Population Data." *Genetics* 63 (1969): 949–67.

———. "The Relation Between Fitness Components and Population Prediction in Orosophila." Parts 1, 2. *Genetics* 68 (1971): 127–49, 151–67.

Ravin, Arnold. "Human Guidance of Human Evolution: Possibilities and Problems." *University of Chicago Magazine* 64 (1972): 9–15.

Reich, Robert, ed. *The Power of Public Ideas.* Cambridge, Mass.: Ballinger, 1988.

Rensch, Bernhard. *Biophilosophy.* New York: Columbia University Press, 1970.

Renshon, S. A. "Assumptive Frameworks in Political Socialization Theory." In *Handbook of Political Socialization*, ed. S. A. Renshon, 3–44. New York: Free Press, 1977.

Restak, Richard. *The Brain: The Last Frontier.* New York: Doubleday, 1979.

Reynolds, Vernon. *The Biology of Human Action.* San Francisco: W. H. Freeman, 1976.

Riley, Matilda White. "Aging, Social Change, and the Power of Ideas." *Daedalus* 107 (Fall 1978): 39–53.

Rintala, Marvin. "Generations: Political Generations." In *International Encyclopedia of the Social Sciences.* Vol. 6, ed. David Sills, 42–46. New York: Macmillan, 1968.

Rosenberg, A. *Sociobiology and the Preemption of Science.* Baltimore: Johns Hopkins University Press, 1980.

———. "On the Propensity Interpretation of Fitness." *Philosophy of Science* 49 (1982): 268–77.

Rozental, S., ed. *Niels Bohr: His Life and Work as Seen by His Friends and Colleagues.* New York: John Wiley, 1967.

Rudner, Richard. *Philosophy of Social Science.* Englewood Cliffs, N.J.: Prentice-Hall, 1965.

Ruse, M. *The Philosophy of Biology.* London: Hutchinson University Library, 1973.

Ryle, G. *The Concept of Mind.* London: Hutchinson and Co., 1949.

Sapir, Edward. *Culture, Language and Personality.* Berkeley: University of California Press, 1956.

Sartori, Giovanni. "From the Sociology of Politics to Political Sociology." In *Politics and the Social Sciences*, ed. Seymour Martin Lipset, 65–100. New York: Oxford University Press, 1969.

Scarr, S., and K. McCartney. "How People Make Their Own Environments: A Theory of Genotype-Environment Effects." *Child Development* 54 (1983): 424–35.

Schlesinger, Arthur, Jr. *A Thousand Days.* Greenwich, Conn.: Fawcett Books, 1967.

Schmitt, Francis O. Foreword. In *The Neurosciences: A Study Program*, ed. G. C. Wharton, T. Melnechuk, and Francis O. Schmitt. New York: Rockefeller University Press, 1967.

Schmitt, Francis O., Stephanie J. Bird, and Floyd E. Bloom, eds. *Molecular Genetics: Neuroscience.* New York: Raven Books, 1982.

Schubert, Glendon. "Biopolitical Behavior: The Nature of the Political Animal." *Polity* 6 (1973): 240–75.

———. "Comment on I. Ebel-Eibesfeldt, 'Human Ethology: Concepts and Implications for Human Ethology.' " *The Behavioral and Brain Sciences* 2 (1979): 1–57.

———. "Brain Science and Political Thinking." Paper presented at the Meeting of the International Society for Political Psychology, Mannheim, Germany, 1981a.

———. "The Sociobiology of Political Behavior." In *Sociobiology and Human Politics*, ed. Elliott White, 193–238. Lexington, Mass.: Lexington Books, 1981b.

———. "Some Implications of Brain Science for Political Science." Paper presented at the Annual Meeting of the Western Political Science Association, Denver, Colorado, 1981c.

———. "Political Ethology." *Micropolitics* 2 (1982): 51–86.

———. "The Evolution of Political Science: Paradigms of Physics, Biology and Politics." *Politics and the Life Sciences*. Vol. 1, 1983, 97–110.

———. *Evolutionary Politics*. Carbondale: Southern Illinois University Press, 1989.

Schull, J. "Are Species Intelligent?" *Behavioral and Brain Sciences* 13 (1990): 63–108.

Scott, John Paul. *The Process of Primary Socialization in Canine and Human Infants*. Monographs of the Society for Research in Child Development, 28 (1), 1963. Lafayette, Ind.: Child Development Publications of the Society for Research in Child Development.

Scriven, Michael. "Explanation and Prediction in Evolutionary Theory." *Science* 130 (1959): 477–82.

———. "Explanations, Predictions and Laws." In *Minnesota Studies of the Philosophy of Sciences*. Vol. 3, ed. Herbert Feigl and Grover Maxwell, 173–90. Minneapolis: University of Minnesota Press, 1962.

———. "New Issues in the Logic of Explanation." In *Philosophy and History*, ed. Sidney Hook, 339–61. New York: New York University Press, 1963.

Searle, John R. "Sociobiology and the Explanation of Behavior." In *Sociobiology and Human Nature*, ed. Michael Gregory, Anita Silvers, and Diane Sutch, 164–82. San Francisco: Jossey-Bass, 1978.

Sears, David. "Political Socialization." In *Handbook of Political Science*. Vol. 2: *Micropolitical Theory*, ed. F. Greenstein and N. Polsby, 93–153. Reading, Mass.: Addison Wesley, 1975.

Sestanovich, S. "Inventing the Soviet National Interest." *The National Interest* (Summer 1990).

Sherrington, Charles S. *Man on His Nature*. Cambridge: Cambridge University Press, 1963.

Sibley, Mulford. "The Limitations of Behavioralism." In *Contemporary Political Analysis*, ed. J. C. Charlesworth, 51–72. New York: Free Press, 1967.

Sidey, Hugh. "History on His Shoulder." *Time*, 8 November 1982, 26.

Sigel, R. S., and M. B. Hoskin. "Perspectives on Adult Political Socialization—Areas of Research." In *Handbook of Political Socialization*, ed. S. A. Renshon, 259–93. New York: Free Press, 1977.

Silvert, Kalman H. *Man's Power*. New York: Viking Press, 1970.

Simpson, George Gaylord. *This View of Life: The World of an Evolutionist*. New York: Harcourt, Brace and World, 1964.

———. *Biology and Man*. New York: Harcourt, Brace and World, 1969.

———. *The Meaning of Evolution*. New York: Bantam Books, 1971.

Sinnott, Edmund W. "The Biological Basis of Democracy." *Yale Review* 35 (1945): 61–73.

Sinsheimer, Robert. "The Brain of Pooh: An Essay on the Limits of Mind." *American Scientist* 59 (January-February 1971): 20–28.

Skinner, B. F. *Science and Human Behavior*. New York: Free Press, 1953.

———. *Beyond Freedom and Dignity*. New York: Bantam Books, 1971.

———. *About Behaviorism*. New York: Vintage, 1976.

———. *The Shaping of a Behaviorist*. New York: Alfred A. Knopf, 1979.

———. *A Matter of Consequence*. New York: Alfred A. Knopf, 1983.

Skolimowski, Henry. "Problems of Rationality in Biology." In *Studies in the Philosophy of Biology*, ed. F. J. Ayala and Theodosius Dobzhansky, 205–24. Berkeley and Los Angeles: University of California Press, 1974.

Smith, B., J. Bruner, and R. White. *Opinions and Personality*. New York: John Wiley, 1953.

Sober, E., and Richard C. Lewontin. "Artifact, Cause and Genic Selection." *Philosophy of Science* 49 (1982): 157–80.

Solzhenitsyn, Alexander. *The Gulag Archipelago, 1918–1956*. Part 1. New York: Harper and Row, 1974.

———. "The Writer Underground." *New York Times Book Review* 3, February 10, 1980, 28–29.

Somit, Albert. "Toward a More Biologically-Oriented Political Science: Ethnology and Psychopharmacology." *Midwest Journal of Political Science* 12 (1968): 550–68.

———. "Biopolitics." *British Journal of Political Sciences* 2 (April 1972): 209–38.

———, ed. *Biology and Politics*. Paris: Mouton, 1976a.

———. "Review Article—Biopolitics." In *Biology and Politics*, ed. Albert Somit, 293–324. Paris: Mouton, 1976b.

———. "Human Nature as the Central Issue in Political Philosophy." In *Sociobiology and Human Politics*, ed. Elliott White, 167–80. Lexington, Mass.: Lexington Books, 1981.

Somit, Albert, Steven A. Peterson, William P. Richardson, and David S. Goldfischer. *The Literature of Biopolitics*. DeKalb, Ill.: Center for Biopolitical Research, 1980.

Sorenson, Theodore Chaikin. *Kennedy*. New York: Bantam Books, 1966.

Sperry, R. W. "Mind, Brain and Humanist Values." In *New View of the Nature of Man*, ed. J. R. Platt, 71–96. Chicago: University of Chicago Press, 1965.

———. "A Modified Concept of Consciousness." *Psychological Review* 76 (1969): 532–36.

———. "Changing Concepts of Mind." In *Man and the Biological Revolution*, ed. R. H. Haynes, 47–61. Toronto: Toronto University Press, 1976a.

———. "Mental Phenomena as Causal Determinants in Brain Function." In *Consciousness and the Brain*, ed. G. C. Globus, G. Maxwell, and I. Savodnik, 163–78. New York: Plenum Press, 1976b.

———. "Mind-Brain Interaction: Mentalism, Yes; Dualism, No." *Neuroscience* 5 (1980): 195–206.

———. "Changing Priorities." *Annual Review of Neuroscience* 4 (1981): 1–15.

———. "Some Effects of Disconnecting for the Cerebral Hemispheres." *Science* 217 (1982): 1223–26.

———. *Science and Moral Priority.* New York: Columbia University Press, 1983.

———. "The New Mentalist Paradigm and Ultimate Concern." *Perspectives in Biology and Medicine* 29, no. 3, Part 1 (1986): 413–22.

Stent, Gunther. *The Coming of the Gold Age: A View of the End of Progress.* Garden City, N.Y.: Natural History Press, 1969.

———. "DNA." *Daedalus* (Fall 1970): 909–37.

———. "An Ode to Objectivity: Does God Play at Dice?" *Atlantic Monthly* 228 (November 1971), 125–30.

———. "Prematurity and Uniqueness in Scientific Discovery." *Scientific American* 227 (December 1972), 84–93.

———. "The New Biology: Decline of the Baconian Creed." In *Encyclopaedia Britannica, The Great Ideas Today*, 152–93. Chicago: Encyclopaedia Britannica, 1976.

———. "You Can take the Ethnics Out of Altruism But You Can't Take the Altruism Out of Ethnics." *Hastings Center Report* 7 (December 1977), 33–36.

Sternberg, R. J. *Beyond IQ.* Cambridge: Cambridge University Press, 1985.

———. *The Triarchic Mind.* New York: Viking, 1988.

Stoufe, I. A. "The Coherence of Individual Development." *American Psychologist* 34 (1979): 834–52.

Stouffer, S. *Communism, Conformity and Civil Liberties.* Glencoe, Ill.: Free Press, 1955.

Strauss, Leo. "Natural Law." In *International Encyclopedia of the Social Sciences*, vol. 11, ed. David Sills, 80–85. New York: Macmillan, 1960.

Stryker, S. "Social Psychology." *American Behavioral Science* 24 (1981): 386–406.

Symons, Donald. *The Evolution of Human Sexuality.* New York: Oxford University Press, 1979.

Tanner, J. M. "Sequence, Tempo, and Individual Variation in the Growth and Development of Boys and Girls Aged Twelve and Sixteen." *Daedalus* 100 (1971): 907–30.

Tatus, Michael. *Mikhail Gorbachev: The Origins of Perestroika.* New York: Columbia University Press, 1991.

Tennov, Dorothy. *Love and Limerance: The Experience of Being in Love.* Chelsea, Mich.: Scarborough House, 1989.

Teylor, Timothy J. *A Primer of Psychobiology.* San Francisco: W. H. Freeman, 1975.

———, ed. *Brain and Learning.* Stamford, Conn.: Greylock, 1978.

Thomas, Alexander. "Behavioral Individuality in Childhood." In *Human Behavior Genetics*, ed. Arnold R. Kaplan, 151–63. Springfield, Ill.: Charles C. Thomas, 1976.

Thomas, L. "Biomedical Science and Human Health: The Long-Range Prospect." *Daedalus* 106 (1977): 163–71.

Thomson, K. S. "Essay Review: The Relationship Between Development and

Evolution." In *Oxford Surveys in Evolutionary Biology*, vol. 2, ed. R. Dawkins and M. Ridley, 220–334. Oxford: Oxford University Press, 1985.

Thornhill, R. "Review of R. W. Matthews and J. R. Matthews's *Insect Behavior*." *Quarterly Review of Biology* 54 (1975): 365–66.

Thorson, Thomas. *Biopolitics*. New York: Holt, Rinehart, and Winston, 1970.

Thurow, L. "Economics 1977." *Daedalus* 106 (1977): 79–94.

Tiger, Lionel, and Robin Fox. *The Imperial Animal*. New York: Delta, 1971.

Toulmin, Stephen. *Foresight and Understanding*. Bloomington: Indiana University Press, 1961.

———. *Human Understanding*. Vol. 1. Princeton: Princeton University Press, 1972.

———. "From Form to Function: Philosophy and History of Science in the 1950's and Now." *Daedalus* 106 (1977): 143–62.

Trivers, Robert. "Parent-Offspring Conflict." *American Zoologist* 14 (1974): 249–64.

———. Contributions to symposium as discussant. In *Human Diversity: Its Causes and Social Significance*, ed. B. Davis and P. Flaherty. Cambridge, Mass.: Ballinger, 1976, 57–75.

———. "Sociobiology and Politics." In *Sociobiology and Human Politics*, ed. Elliott White, 1–43. Lexington, Mass.: Lexington Books, 1981.

Tucker, Robert C. *Politics as Leadership*. Columbia: University of Missouri Press, 1981.

Venable, Vernon. *Human Nature: The Marxian View*. New York: Meridian, 1966.

Verba, E. "Evolution, Species and Fossils. How Does Life Evolve?" *South African Journal of Science* 76 (1980): 61–84.

Waddington, Conrad Hal. *The Ethical Animal*. London: Allen and Unwin, 1960.

———. "The Basic Ideas of Biology." In *Toward a Theoretical Biology*, ed. Conrad Hal Waddington, 1–31. Chicago: Aldine, 1968.

———. "Mindless Societies." In *The Sociobiology Debate*, ed. Arthur L. Caplan, 202–58. New York: Harper and Row, 1978.

Wahlke, John. "Pre-Behavioralism in Political Science." *American Political Science Review* 73 (1979): 9–32.

Wald, George. "Determinancy, Individuality, and the Problem of Free Will." In *New Views of the Nature of Man*, ed. John R. Platt, 16–46. Chicago: University of Chicago Press, 1965.

Warren, Howard C. *Dictionary of Psychology*. Boston: Houghton Mifflin, 1934.

Washburn, Sherwood L. "Human Behavior and the Behavior of Other Animals." *American Psychologist* 33 (March 1978): 405–18.

Watson, James O. *The Double Helix*. New York: Signet Books, 1968.

———. "Moving Toward the Clonal Man." *Atlantic Monthly* 227 (May 1971), 50–53.

Watson, John B. *Psychology from the Standpoint of a Behaviorist*. Philadelphia: J. B. Lippincott, 1919.

———. *Behaviorism*. New York: W. W. Norton, 1924.

Weber, Max. *The Theory of Social and Economic Organization*. Glencoe, Ill.: Free Press, 1947.

———. *On the Methodology of the Social Sciences*. Glencoe, Ill.: Free Press, 1949.

Weissberg, Robert. *Political Learning, Political Choice and Democratic Citizenship*. Englewood Cliffs, N.J.: Prentice-Hall, 1972.

Weisskopf, Victor F. "The Frontiers and Limits of Science." *Daedalus* 113 (1984): 177–95.

Westman, Jack, ed. *Individual Differences in Children*. New York: John Wiley, 1973.

White, D. M. "Power and Intentions." *American Political Science Review* 45 (September 1971): 749–59.

White, Elliott. "Intelligence and the Sense of Political Efficacy in Children." *Journal of Politics* 3 (1968): 710–31.

———. "Intelligence, Individual Differences and Learning: An Approach to Political Socialization." *British Journal of Sociology* 20 (1969): 50–66.

———. "Genetic Diversity and Political Life: Toward a Populational-Interaction Paradigm." *Journal of Politics* 34 (November 1972): 1203–42.

———. "Sociobiology and Politics." *Political Science Reviewer* 8 (1978): 263–86.

———. "Genetic Diversity and Democratic Theory." In *Through the Looking Glass: Epistemology and the Study of Political Inquiry*, ed. Maria Falco, 311–28. New York: University Press of America, 1979.

———. "Clouds, Clocks, Brains and Political Learning." Paper presented at the Meeting of the International Society for Political Psychology, Boston, Massachusetts, June 1980.

———. "The Neurobiological Basis of Human Action." Paper presented at the Annual Meeting of the Western Political Science Association, Denver, Colorado, 1981a.

———. "Political Socialization from the Perspective of Generational and Evolutionary Change." In *Sociobiology and Human Politics*, ed. Elliott White, 259–81. Lexington, Mass.: Lexington Books, 1981b.

———. "Sociobiology, Neurobiology, and Political Socialization." *Micropolitics* 1, no. 2 (1981c): 113–44.

———. "Brain Science and the Emergence of Neuropolitics." *Politics and the Life Sciences* 2, no. 1 (1982a): 23–25.

———. "Self-Direction and Political Action: Conscious Purpose, Emic Analysis, and Ongoing Political Biography." Paper presented at the Annual Meeting of the Northeastern Political Science Association, New Haven, Connecticut, November 18–20, 1982b.

———. "Review of *Promethean Fire*." *Politics and the Life Sciences*, 1984.

———. "Brains, Bonds and Bureaucracy." In *Biology and Bureaucracy*, ed. E. White and J. Losco. Lanham, Md.: University Press of America, 1986, 276–313.

———. "Shadow Networks and Self Selection: The Neuropolitics of Covert Action." Paper prepared for delivery at the Annual Meeting of the American Political Science Association, San Francisco, 1990.

———. "Politics of the Master(s) Art?" *Politics and the Life Sciences* 9 (1991): 71–73.

———. "Locals, Cosmopolitans and Politics: The American Founders from a Neuropolitical Perspective." In *Human Nature and Politics*, ed. J. Losco, Forthcoming.

White, Leslie. *The Science of Culture*. New York: Grove Press, 1949.

Whitehead, Alfred North. *Dialogues*. New York: Mentor Books, 1956.

———. *Adventures in Ideas*. New York: Mentor Books, 1958.

———. *Science and the Modern World*. New York: Mentor Books, 1968.

Wicklynd, R. A. "The Influence of Self-Awareness on Human Behavior." *American Scientist* 67 (1979): 187–93.

Wiener, Norbert. *Cybernetics*. Cambridge, Mass.: MIT Press, 1961.

———. *God and Golem, Inc.* Cambridge, Mass.: MIT Press, 1964.

Wigner, Eugene P. "The Limits of Science." In *Readings in the Philosophy of Science*, ed. H. Feigl and M. Brodbeck, 757–66. New York: Appleton-Century-Crofts, 1953.

Williams, George C. *Adaptation and Natural Selection*. Princeton: Princeton University Press, 1966.

Williams, M. "The Logical Status of the Theory of Natural Selection and Other Evolutionary Controversies." In *The Methodological Unity of Science*, ed. M. Burge, 84–102. Oasdrecht: Reidel, 1973.

Williams, Roger J. *Biochemical Individuality*. New York: John Wiley, 1956.

———. *Nutrition Against Disease*. New York: Bantam Books, 1973.

Wills, Richard H. *The Institutionalized Severely Retarded*. Springfield, Ill.: Charles C. Thomas, 1973.

Wilson, Edward O. *Sociobiology: The New Synthesis*. Cambridge: Harvard University Press, 1975.

———. "Biology and the Social Sciences." *Daedalus* 106 (Fall 1977a): 127–40.

———. Foreword. In *Sociobiology and Behavior*, by David P. Barash, xii-xv. New York: Elsevier, 1977b.

———. *On Human Nature*. Cambridge: Harvard University Press, 1978.

———, et al. *Life, Cells, Organisms, Populations*. Sunderland, Mass.: Sinauer Associates, 1977.

Wilson, P. S. *Man, the Promising Primate*. New Haven: Yale University Press, 1980.

Wilson, Woodrow. "The Law and the Facts." *The American Political Science Review* 5 (1911).

Wiseman, H. Victor. *Politics: The Master Science*. New York: Pegasus Press, 1969.

Wrangham, R. W. "Sociobiology: Modification with Dissent." *British Journal of the Linnean Society* 13 (1980): 171–217.

Wright, Sewall. *Evolution and the Genetics of Populations*. Vol. 4, *Variability Within and Among Natural Populations*. Chicago: University of Chicago Press, 1978.

Young, J. Z. *Programs of the Brain*. London: Oxford University Press, 1978.

———. *Philosophy and the Brain*. New York: Oxford University Press, 1987.

Index

About the Author

ELLIOTT WHITE is a Professor of Political Science at Temple University, Tokyo campus. He received his Ph.D. from the University of Chicago. Dr. White has edited *Sociobiology and Human Politics* (1981) and co-edited *Biology and Bureaucracy* (1986).

www.ingramcontent.com/pod-product-compliance
Lightning Source LLC
Chambersburg PA
CBHW031953180326
41458CB00006B/1704